ECOTHEOLOGY AND THE PRACTICE OF HOPE

SUNY SERIES ON RELIGION AND THE ENVIRONMENT
Harold Coward, editor

Ecotheology
and the
Practice
of
Hope

ANNE MARIE DALTON
AND
HENRY C. SIMMONS

Published by State University of New York Press, Albany

© 2010 State University of New York

All rights reserved

Printed in the United States of America

No part of this book may be used or reproduced in any manner whatsoever without written permission. No part of this book may be stored in a retrieval system or transmitted in any form or by any means including electronic, electrostatic, magnetic tape, mechanical, photocopying, recording, or otherwise without the prior permission in writing of the publisher.

For information, contact State University of New York Press, Albany, NY
www.sunypress.edu

Production by Diane Ganeles
Marketing by Michael Campochiaro

Library of Congress Cataloging-in-Publication Data

Dalton, Anne Marie.
  Ecotheology and the practice of hope / Anne Marie Dalton and Henry C. Simmons.
       p. cm. — (SUNY series on religion and the environment)
  Includes bibliographical references (p.   ) and index.
  ISBN 978-1-4384-3297-7 (hardcover : alk. paper)
  ISBN 978-1-4384-3296-0 (paperback : alk. paper)
 1. Ecotheology. 2. Knowledge, Sociology of. I. Simmons, Henry C. II. Title.
BT695.5.D35 2010
261.8'8—dc22

                                                             2010007172

10 9 8 7 6 5 4 3 2 1

Contents

	Introduction	vii
CHAPTER 1	The Social Imaginary and the Ecological Crisis	1
CHAPTER 2	The Emergence of Ecotheology	19
CHAPTER 3	Imagined Futures	39
CHAPTER 4	Theology and the Ecological Crisis	53
CHAPTER 5	Science and Ecology	71
CHAPTER 6	Global and Local in the Social Imaginary	89
CHAPTER 7	Living *As If*	105
	Conclusion	125
	Notes	127
	Bibliography	155
	Index	181

Introduction

This is a book that puts together two realities that seem increasingly in tension—the ecological crisis and hope. This is so because the ecological crisis continues to seep into our beings, producing anxiety, sadness, guilt, and sometimes despair. The world seems, as David Rutledge says of his own feelings, "out of control."[1] A similar deep anxiety sets Christopher Turner on a voyage to find signs of hope for a future for his young daughter.[2] Although the hope we seek in the light of ecological devastation almost always seems elusive, it remains the motivational horizon against which many find the energy to confront the crises.

The word hope can suggest a cheerful whistling in the dark optimism. But it can also have the hard edge of the concrete realities we face—and in the face of these realities, hope can actively resist despair and effect concrete changes in how we conduct our lives. There exists considerable theological literature on the virtue of hope within the Christian tradition.[3] It is not our intention in this book to explore the theological meanings of hope. Rather our purpose is to raise up the evidence of hope, specifically the practices that inhere in the texts of Christian theologians[4] using the tools of their scholarly training as well as their own faith to confront the ecological crisis. This sense of hope is akin to Bill McKibben's, who summarizes the impact of his exposure to alternative ways of living than that of the dominant North American society when he writes, "I found proof there that there are less damaging ways to lead satisfying human lives, evidence that infatuation with accumulation and expansion is not the only possibility. . . . Real hope implies real willingness to change perhaps in ways suggested by this volume."[5]

Our volume speaks to hope as willingness to change at all levels of human life because at all levels of practice the interweaving of vision and action is at work constructing and reconstructing the contours and norms of human attitudes and behaviors within a particular society.

Practices of hope resist despair even in the face of evident ecological degradation. As practitioners of hope, Christian theologians are sober in their judgments about the certain results of business as usual in our relationship to the earth; still, they do not give up their intense efforts to pull the world back from the brink of ecological disaster. This book deals with the work of theologians of the past 50 years who self-identify as Christian, and who have produced books and articles whose clear intent is to bring Christian resources to bear on the ecological crisis. Collectively and individually, the works of these Christian scholars of theology, Bible, ethics, and religious education—ecotheologians, to use the common shorthand designation—are practices of hope.

We call these works of ecotheology texts because the word texts connotes writings that have a life beyond the author's intent and so engage the reader in a process of interpretation. They rely on the power of language. Many scholars, in particular of the late twentieth century, have attempted to articulate the integral place of language in the very construction of humanity and indeed of whatever it is we call reality.[6] Our argument for the power of ecotheological texts relies in large part on this scholarship. Wilhelm von Humboldt referred to the emergence of language as "the moment of connection between nature and idea."[7] It is in this sense that we understand the ecotheological texts we examine as serious instances of efforts to reestablish that original moment of connection. Furthermore, it is our judgment that these texts with their horizon of transformation participate in a social desire that in Chaia Heller's understanding overcomes an inherited dualism between feeling and thinking, between the social and the individual, and between romanticism about nature and activism on behalf of creation, to name a few.[8] They stir not only new thoughts, but new imagination and new impetus for further action, in a word, new hope.

The ecological crisis provokes both grief and hope. We grieve for beauty that has passed and will continue to pass away. We hope that a beauty will still remain. The pre-Socratic Greek philosopher Herakleitos, who taught that the universe is in constant change and that there is an underlying order or reason to this change, said, "The things of which there is seeing, hearing and perception, these do I prefer."[9] This saying speaks to the critical relationship of human life to grief and hope experienced in the "ten thousand things" of the *Tao Te Ching*: "Nameless: the origin of heaven and earth. Naming: the mother of ten thousand things."[10] Our very humanness is exercised in relation to "the things of which there is seeing, hearing, and perception," these "ten thousand things." If the beautiful things of the earth disappear, how will humans know their own humanness?

The ecological future of the planet readily provokes grief that can edge on despair, precisely because of the ecological disasters of which we are collectively increasingly aware—global warming, the decimation of rain forests, the endless production of unrecyclable garbage, toxic waste, diminishing biodiversity, accelerating consumption of nonrenewable resources, and widespread disease and starvation. While there are also many appeals to hope based on the inherent goodness of humankind or the promises of religious faith or the visible efforts of groups and individuals, the evidence for hope may look weak and ineffectual against the overwhelming reasons for grief. We ourselves experienced this.

The original shape of this book started out in quite a different direction from the present text. Briefly, it was an attempt to organize the writings of ecotheologians and eco-ethicists in a way that would facilitate conversation between individuals and groups who came to ecotheology and eco-ethics from different perspectives: ecology as political text, gender text, theological text, institutional text, international text, and so on.[11] A year into the project we were spiraling into a dark place. All the texts pointed to impending disaster. It was in reading Charles Taylor's *Modern Social Imaginaries* that we came to an insight that pulled us back from despair: as dark a picture as ecotheologians painted, they kept writing, lecturing, and researching, some for 30 or 40 years or more. The repetition and growth of these ideas, the ongoing development of groups of colleagues and "disciples," and the relentless push to give ecotheology and eco-ethics a legitimate place in the academy have been subtly shifting the interplay of understandings and practices that have broadly shared moral legitimacy, or, stated more simply, are the way we live our lives together.[12] It is this that Charles Taylor describes as the social imaginary.

With this insight, we abandoned our original project and moved to the present one, *Ecotheology and the Practice of Hope*. We have experienced the power of texts that shape the social imaginary, that make legitimate the compelling interests of the earth, that give moral status to particular concerns for the environment, and that can in doing this lift people from a position of near despair to resisting despair—to hope.

This story illustrates what we understand by practices of hope. It is in the murkiness of everyday life that one searches for and finds practices of hope that change the way the world does business. This has been the life task of the theologians and ethicists whose texts are a source of hope.

The hope generated by the texts we consider here is not foolishly optimistic or unfounded. The situation is grave, and the authors recognize it; there is no guarantee that the drift toward an ecological nightmare can be halted. Nor do we propose a new world order, a utopian hope that the nonecological ways of Western social modernity can be overturned. There is no evidence that such a social revolution is likely or possible. At a psychological level, our understanding of hope is close to what a colleague, Gabriel Moran, calls "resisting despair." For 50 years Christian ecotheologians have seen with clarity the depth and breadth of ecological degradation. These are people who have looked into the abyss and asked if it is too late to save the planet. At the theological level, that is, at the level at which these people work professionally, the very act of continuing to expose the problems, formulate strategies for collective and communal action, articulate new understandings that make Christian faith more clearly compatible with the needs of the planet, is an act of resistance against despair—a practice of hope.

These ecotheologians are at one level idealists—they desperately want the earth to be healed. But, as one reads the small signs of the times that they cite as giving them hope to continue to work for change, it becomes clear that they are realists—they know change will be negotiated, will be incremental, and will happen by small modifications of the existing social imaginary. On what basis must Christian scholars continue to do this work and why do we cautiously hold that the work of Christian scholars matters for real change?

First, the texts we read are engaged texts in the sense that Elizabeth Johnson understands this term, that is, they generate "genuine transformation . . . to read our current situation honestly, think long and hard about how to make changes that would foster healthier societies, and identify and support promising trends that . . . help get our societies moving in the right direction."[13] These engaged texts are themselves practices of hope. While a facile or even common reading may see these texts as expressions of understanding and explanation only, we are arguing that within an understanding of the construction of a society and culture, such as that represented by Taylor's notion of the social imaginary, these texts participate in an engaged community of scholars and activists and so function as practices of a hope for the transformation of societies.[14] They are situated within the power of language especially when constructed as text to inflame the imagination, supply new language, and renovate the worldview. Thus they find further expression in multiple activities from individual lifestyle changes to political resistance.

The second reason to assert that Christian scholarship in religion and ecology taken collectively is indeed capable of effecting change is tied to a complex interweave that sociologists call the social imaginary, which allows us to see how environmental change can happen. The social imaginary is how we imagine our lives together and how that imagination gives rise to the practices of our lives. The social imaginary is the interplay of understandings and practices that have broadly shared moral legitimacy. And the social imaginary is malleable.

What we grieve and hope for, what is seen to be important, is part of our social imaginary. But what we grieve and hope for can change over time. Indeed, it has been shown historically that the social imaginary is a dynamic process that has changed over time and can be expected to continue to change. In the social imaginary of the West, understandings and practices that once had shared moral legitimacy have changed: emancipation, women's suffrage, aboriginal rights, child labor laws, civil rights, and so on. From today's vantage point, all these changes may seem inevitable, even self-evident. But in each historical context the dogged persistence of the few who fought for these causes seemed quixotic and highly unlikely to change the way in which people imagined how the world should be.

The social imaginary does change, although not easily. Even in the face of ecological crisis, this process of change offers a basis for hope for the "things of which there is seeing, hearing and perception." An ecologically responsible world is possible, and the engagement of Christian scholars in imagining that future plays a critical role in bringing it about. This is not naive or cheerful optimism. As this book will show, it is a tenacious hope grounded in a sophisticated analysis of what already is happening and an understanding of how the social imaginary can change.

The texts of theologians seek to create a cultural space. These texts look for a place within a secular world for religious discourse, in this case Christian discourse. Ecotheologians are further concerned to introduce to the Christian community a discourse about ecology and in turn to influence the broader public. They bring the immediacy of the ecological crisis to the Christian community so that it understands how this crisis "feels here and now."

But the goal is more than awareness. Ecotheologians seek transformation in the way human societies function in relation to the rest of the natural world. Ecotheologians do not act alone and their texts do not exist in isolation. While it is possible that single visionary figures such as Thomas Berry or Joseph Sittler may have considerable influence, this is not so for all ecotheologians. Ecotheological texts exist in conversation with each other and with the broader communities of academic theologians, churches, publics, and related disciplines. For that reason we consider the writings of ecotheologians cumulatively, because it is cumulatively that these texts have the best chance of penetrating the present social imaginary and influencing change toward ecologically responsible practices. Thus Christian texts are engaged in a process of re-creating the social imaginary, "the ways people imagine their social existence, how they fit together with others, how things go on between them and their fellow citizens, the expectations that are normally met, and the deeper normative notions and images that underlie these expectations."[15] The overall purpose of this book is to demonstrate that ecotheological texts do engage, promote, and influence change. This is so even in a society where practical, immediate, material, and short-term benefits often determine the popular notions of engagement and change.

We contend that Christian ecological texts like ecological texts in general contribute to the renaming of how things usually go and a reimagining of how they ought to go in the light of the ecological crisis. These texts contain ideas and ideals about a new moral order. They call for the engagement of religious institutions and adherents in the ecological crisis. They pay attention to the manner in which religious images, rhetoric, and concepts function within the modern social imaginary and how they might contribute to directing change in the social imaginary to greater ecological responsibility.

Chapter 1, "The Social Imaginary and the Ecological Crisis," introduces Charles Taylor's account of the social imaginary, a term used to describe "that common understanding that makes possible the common practices and a widely shared sense of legitimacy" that incorporates both factual and normative senses of "how we all fit together."[16] Taylor's exploration of the social imaginary of the West is the framework for our argument that the interplay of new understandings and new meanings participates in critical ways in producing societal change. The social imaginary changes over time as "what is originally just an idealization grows into a complex imaginary through being taken up and associated with social practices, in part traditional ones but ones often transformed by the contact."[17]

Chapter 2, "The Emergence of Ecotheology," examines two contexts in which Christian ecotheological texts emerged—the sociopolitical conversation that heightened ecological awareness in the mid-twentieth century, and the theological conversation that eventually legitimized ecotheology in the academy. These contexts are substantially different.

Chapter 3, "Imagined Futures," examines early Christian ecotheological texts of the mid-twentieth century that expressed the convictions of Christian preachers and scholars—some of whom became prominent ecotheologians and are still identified with the field—that a response to the ecological crisis was urgent. It shows that early personal experiences became sources of the passion for the earth in the texts of Christian ecotheologians. Early ecotheological texts initiated the reformation of Christian teachings about creation and human responsibility for its integrity. They imagined a new kind of future.

Chapter 4, "Theology and the Ecological Crisis," shows how theology frames the ecological crisis. It demonstrates what a Christian theological perspective adds to Taylor's understanding of the social imaginary—faith in a divine dimension to history, a presence of an active and caring God who has already redeemed the world. Indeed, a large part of the work of the ecotheologians is a strenuous reclaiming of the physical world (earth or entire cosmos) as a sphere of God's action and caring.

Chapter 5, "Science and Ecology," explores ways in which Christian writers on ecology use science as a major conversation partner. Scientific understandings and practices have profoundly shaped that social imaginary. Ecotheologians challenge some of the ways in which science has functioned and also engage science in the issues of the environmental crisis.

Chapter 6, "Global and Local in the Social Imaginary," focuses on the impact of context on the work of ecotheologians. In taking location seriously, the specificity of the social imaginary of the West becomes clearer and more delimited and thus changed by attention to the legitimacy of other locations. Location involves many aspects of the ways people are in the world: geopolitical context, socioeconomic status, gender-in-context, race, and communities of meaning.

Chapter 7, "Living *As If*," focuses on practices of hope and a new social imaginary. A new social imaginary comes about when new practices—practices of hope—emerge together with new understandings of humans and the earth, and when ecological practices are imbued with new meaning. The ecotheology we have discussed so far is a practice of hope. In the face of environmental degradation, it arises out of visions of hope, by women and men who resist despair. As a community of scholars they carry forward a cumulative vision, pushing the boundaries of academic disciplines, and inspiring renewed practices in local communities and congregations. In this chapter we look at a sample of these renewed practices in areas from agriculture to worship. We identify these practices as energy and promise for reshaping a social imaginary.

1

The Social Imaginary and the Ecological Crisis

Vast, hideous, and desolate
—William Bradford
on glimpsing the New World.[1]

William Bradford's description of the New World contrasts sharply with that of Thomas Berry. To Bradford the natural wilderness, although well inhabited by North American native peoples, was desperately lacking in the signs of modernity—that is, of human progress; it needed taming, subjugation, and human artifact. To Berry it was the Eden we have since wrecked; in his words: "When we came to this continent, it was a glorious land of woodlands and prairie grasses . . . a land of abundance."[2] This contrast sets in bold terms the question: How did we get to this place in history we now describe as an ecological crisis? The answer, put simply, is Western modernity.[3]

This chapter will first complicate this response but not deny it. It will address the elements in the development to the present Western way of life that have led us to such a poignant moment, in which the future of the planet is at stake. These are elements that will be quite familiar to most of our readers and are certainly present in many of the ecotheological texts we will examine in the chapters ahead. Second, this chapter will propose a framework for a deeper understanding of the nonintentional, even oblivious, way in which the construction of a Western modernity as described by Charles Taylor produced the "wrecks of Eden." Western modernity, however, is not a rigid, static construction. It has been and continues to be malleable. The social imaginary carries its own potential for change, even radical change, and it is here that the hope for a better future, within a Western perspective, at least, can be envisioned and accomplished. Third, this chapter will address the particularity of vision brought by Christian ecotheology texts to the hope for a more ecologically sound social imaginary. Our argument is that as engaged texts they may play a significant role in the shifting of our imagination and practice. For that to happen, however, there must be an understanding of the location of religion within Western modernity.

How did Western modernity as we have constructed it result in such a "wreck of Eden"? Two elements that are reflected in the earlier comment of Bradford are the dominance of a scientific worldview and the shift from a cosmic-centered to a human-centered perspective on the universe. A scientific worldview is not merely a physical science-based view; it refers to the merging of the notion of history as humanly controlled progress with the development of powerful technologies. The virtual eradication of the idea that humans were subject to a divine will or to Fate in some form translated into the further idea that there was virtually no limit to the progress that could be accomplished with the right knowledge and tools. For all of this the natural world was merely the resource. The conviction was that it would only be a matter of time before disease, poverty, and illness would be under control.

The shift from a cosmic-centered to a human-centered perspective is really another thematization of the same reality. The perception that human society ought to fit in some way into a cosmic order as reflected in the seeming predictability and graceful movement of the planets, the seasons, and life and death in the organic world, gave way over time to notions of society that were unrelated to the physical world. The pre-Copernican worldview had the earth at its center. Humans gained their importance from their geographical location on earth as the center of the cosmos. In the post-Copernican view the centrality of planet earth no longer held. Human importance now had to be constructed apart from the earth; most notably humans were seen as the only beings made in the image of God. Human society sat on top of the creation; creation was a mere backdrop for the important work of building a better future, whether this was to end in this world or in some supernatural world. These broad strokes were not obvious overnight, but can be seen only in retrospect. Various scholars, including those whose texts we consider in later chapters, have focused on the components of these historical shifts. Taylor considers them within his description of the construction of a social imaginary that is our present Western modernity.

THE SOCIAL IMAGINARY OF THE WEST

Taylor uses the notion of the social imaginary, which originated in the social sciences, to provide a framework for understanding how we got to be where we are, and also for understanding how societies can change.

Thus, there is not only one social imaginary, but rather there are multiple social imaginaries, in fact, as many as there are recognizable actual and potential societies.[4] Taylor's concept of social imaginaries emerges from an historical discussion about the construction of human communities, which has deep roots in Western scholarship.[5] In particular, Benedict Anderson's notion of imagined communities helps to locate Taylor's notion of social imaginaries within that discussion. Anderson's explanation of how a nation is imagined is most relevant to our use of Taylor's concept in that it elucidates the nuanced meanings of "imaginary."[6] According to Anderson, nations are *imagined* because they consist of members we will never meet or know. Yet we imagine ourselves in some sort of communion with them all. Thus, nations are invented not awakened. As a departure from previous scholars' understanding of "invention," Anderson associates invention (as does Taylor) not with fabrication and falsity, but with imagining and creating.[7] The nation is imagined as limited; it has set boundaries, even if elastic, outside of which lie other nations. It is imagined as sovereign; the idea of nation was conceived at a time when the notion of a divinely ordained authority was losing legitimacy. The nation, finally, is also imagined as a community; a fraternity or comradeship to which one is committed enough to lay down one's life.

These notions of the nation as imagined community as well as Anderson's work on its cultural roots are brought forward into Taylor's conception of a social imaginary. What Taylor adds is the focus on the interweave of theory and practices that constitute the dynamics of the imaginative process by which one creates a society. Furthermore in using Western modernity as an instance of social imaginary at work, he focuses on the very theories and practices critiqued by ecotheologians of late twentieth century as coalescing in an ecologically unsustainable society. On the other hand, by articulating the apparatus by which one constantly constructs and reconstructs a social imaginary, Taylor offers (intentionally or not) a potential means by which even this ecologically indifferent or even hostile social imaginary of the West can be redirected. According to Taylor, the social imaginary is "not a set of ideas, rather, it is what enables, through making sense of, the practices of a society."[8] The term carries the notion that societies develop as a result of the application of imagined futures.[9] Evidence that sustained effort from different starting points over time can effect change in the social imaginary forms the basis for hope. The community of scholarship on religion and ecology of the last several decades has already helped shape a more ecologically sensitive social imaginary. It is different from but also identifiable within the present Western social imaginary, which has virtually excluded ecological concerns. It represents a redirection of the manner of human life based on a new imagination. Yet, it rests on the basic principle that ideas and practice in the service of powerful convictions and imagination regarding a future are effective over time.

"Central to Western modernity," Taylor argues, "is a new conception of the moral order of society."[10] Taylor is speaking of the emergence of a set of ideas and practices in social forms that characterize modernity in the West. These include, preeminently but not exclusively, the market economy, the public sphere, and the self-governing people.[11] The social imaginary, then, refers to the way in which people imagine their manner of living together. The source of the particular imagined existence may begin with the articulation of certain ideas, but it is the images, stories, and legends that grow out of or mutate from these ideas that form the social imaginary. While a theoretical understanding is often clearly grasped only by a few, many share the social imaginary. The social imaginary is "that common understanding that makes possible the common practices and a widely shared sense of legitimacy."[12]

Furthermore, the social imaginary is both factual and normative. It is at the same time recognition of how things usually work and how they ought to work. The social imaginary of a particular modernity or a given society governs all social behaviors, such as how to conduct oneself at a social gathering or how to negotiate the fair treatment of competing interests. While the concept of the social imaginary may seem abstract, it is immediate, practical, and essential for society. For example, it legitimizes a particular school curriculum, adequate school behaviors, tests and standards, and anticipated outcomes (including acceptable retention and graduation rates).

The social imaginary cannot be understood only in terms of an *articulated* history. Besides the existence of ideas and events that led to the present imaginary there is a large and *inarticulate* understanding, a "wider grasp" of such questions as "how we stand to each other, how we got to be where we are, how we relate to other groups and so on."[13] As Taylor explains,

> This wider grasp has no clear limits. . . . It is in fact that largely unstructured and inarticulate understanding of our whole situation, within which particular features of our world show up for us in the sense they have. It can never be adequately expressed in the form of explicit doctrines because of its unlimited and indefinite nature. That is another reason for speaking here of an imaginary and not a theory.[14]

In creating and sustaining the social imaginary, practices and their meanings work together. Understanding often gives rise to practice, but practice itself carries an understanding. Usually a common understanding precedes any kind of theory about why those involved might act in this way. Thus, for instance, we all know how to operate toward each other within different social spaces, how to respect boundaries of a political demonstration, how to organize and carry out an election.[15] A striking example of this is the enduring anger of many Americans towards the Supreme Court for its decision favoring Bush over Gore in the American presidential election of 2002.

While largely recognizable as continuous over time, practices and their meanings change, sometimes slowly and organically, but sometimes quickly and abruptly. When practices begin to change and boundaries are consistently transgressed, we are probably in transition. Idealizations, theories, and ideas grow into a complex imaginary as they are incarnated in new practices or associated with old ones that are transformed by that process.[16] Examples include the ways in which ideas of Karl Marx or Adam Smith inform present practices in Western societies. The original ideas may be barely recognizable in important segments of societies, because of creativity and practical applications (as well as human self-interest and greed). Likewise, the visions and ideas of the Christian reformers of the sixteenth century are still recognizable today in mainline Christian churches. These reformers, however, would likely be shocked at particular practices and views to which their visions have given rise.

It is clear, then, that the social imaginary of a group is not merely the fruit of a well-reasoned theory, a vision of the imagination, or useful practices. It is rather the combination of these as they enter the process of human living and the struggle to fulfill both the basic and higher needs and desires that govern life together.

In the West, this combination has come to produce what is commonly called Western modernity. It is not the only possible modernity. Different sets of ideas, visions, and practices or other ways of relating the theories and practices could have produced a quite different society. What we have is the contingent way in which history shaped our present modernity in the West. It is in this modernity that the ecological crisis emerged and must be confronted. The key forms, as Taylor describes them, all hold critical difficulties from an ecological perspective. But the dynamics of the social imaginary itself also hold a promise that a different, more ecological, way of life is possible.

Key Forms in the Social Imaginary of the West

Taylor argues that the moral order that shapes the social imaginary of Western modernity is based on the concepts of individual rights and mutual benefit. Beginning in the seventeenth century, the notions that society was constituted by rational beings and that governing powers could be challenged began to take hold. Gradually these ideas infiltrated various segments of the society—classes, races, genders, and ethnic groups. They shaped and were shaped by three main cultural forms: the economy, democratic self-rule, and the public sphere.[17] Each of these forms has significant implications for the subsequent ecological crisis. The specificity of the forms and the dynamic manner in which they arose, however, demonstrate both the malleability of the social imaginary and the possibility of alternatives as practices are transformed or as new ones emerge.

Taylor explains the sources of these three main cultural forms in a much more complex way than we can here.[18] It is not our purpose to give a full account of the emergence of the modern social imaginary as he does. Rather we give a somewhat schematic account to illustrate the manner in which ideas and practice interweave; that is, how they are imagined and incorporated in the formation of an ever-changing social imaginary.

The Economy

The specific capitalist form of the economy in the West originates with ideas of Adam Smith (at least they are considered to be best articulated by him). Briefly, what is significant for our purposes is the way in which the relationships among creatures came to be redefined. Prior to the theories Adam Smith articulated, the hierarchical distribution of power in the human world was believed to originate in the order of the cosmos itself. What emerged was the understanding that the source of power lay within the human world.

It is within this context that Smith articulated his economic theory. He assumed a world governed by humans for their own purposes, rather than a world in which the distribution of power was received from divine or cosmic sources. As they gradually interacted with and changed practice, Smith's theories led to the belief that by concentrating on building one's own wealth, one benefited the whole society through the working of "the invisible hand." Self-interest becomes benevolence. According to Taylor, this kind of profitable exchange for the sake of security and mutual benefit becomes the metaphor for the whole political society.[19] We come to see society itself as an economy, "an interlocking set of activities of production, exchange and consumption, which form a system with its own set of laws and its own dynamic."[20] This is now a system that is separate from the polity. It functions by its own laws, laws that humans need to know in order to live within society. Further, these laws, which in their origins gave specific answers to specific economic questions, now have taken on the character of abstract principles that are immutable and unarguable.[21]

It is worth noting and an illustration of how particular elements of imagined futures change over time that Smith presumed a certain kind of community interest and civic responsibility as part of one's self-interest. So his theories and the manner of their practice in his own day become quite different historically as that sense of community gives way to a much more individualistic society.[22]

The Public Sphere

Although the economy may have been the first to disengage itself, it paved the way for other forms to take on independent existence. However, while the economy depends largely on individual activities, the public sphere and democratic self-governance stem from a new notion of the collectivity and its power and function.

The public sphere relies for its existence on the notion of secularity and its sense of profane time, which is discussed more fully below where we examine the role of religion in the social imaginary of the modern West. The public sphere is not identical to any given or officially constituted government, but rather has power over it. Taylor describes it as that public space in which discussion, via various media, produces a public opinion. Thus it is a metatopical agency. It is metatopical in that it is nonlocal and transcends any one assembly or topic of discussion. It is an agency because it has power, a perception of itself as an entity that stands outside the political order and exerts a certain kind of power over it. For example, public officials answer to the public sphere.[23] Like the social imaginary itself, the public sphere did not result merely from a set of ideas. It came into existence in concrete occasions of assembly and through the growing influence of the print media and print capitalism.[24]

Self-Governance

Taylor explains two examples of the emergence of the self-governing democratic form in the modern social imaginary—the American Revolution and the French Revolution. They exemplify two different ways, one a more recognizable continuity and the other a more radical discontinuity, in which the social imaginary can be substantially changed.

In the case of the American Revolution, the colonists understood themselves as Englishmen (male gender intended, as women were not yet recognized as legal persons). As Englishmen they belonged to England and thereby possessed certain rights that had been institutionalized in governing assemblies in their own land, where parliament had been instituted. The power of the people was represented beside the power of the King. These institutions had already inculcated the idea that peoples could be founded (and refounded) not in some mythical past, but by the people themselves. What became an essentially extended and new idea of the equality of all peoples (not just Englishmen) in the U.S. Constitution relied for its legitimacy on older forms. Legitimacy was transferred not from the historical form represented by the King but by natural law. "Truths held self-evident" refers to the foundation of order and legitimacy within natural law. However, the idea of foundation is transformed into a foundation totally in the will of the people, that is, in elections. In Taylor's words:

> Older forms of legitimacy are colonized, as it were, with new understandings of order, and then transformed, in certain cases, without a clear break.... But what has to take place for this change to come off is a transformed social imaginary, in which the idea of foundation is taken out of the mythical early time and seen as something that people can do today.[25]

At the beginning of the Revolution, Americans saw themselves fighting for their rights as Englishmen; at the end, they had broken with the King and stood for *universal* human rights. It is also clear that the transformation was not a matter of interpretation only. The institutions for the incorporation of this new interpretation, especially that of an elected assembly, were already in existence.[26]

The French Revolution, however, was different. The French had to move away from the idea and practice of legitimate dynastic rule to that of self-governance without any "agreed meaning in a broadly based social imaginary."[27] This was in large part the reason, Taylor claims, for the period of instability that followed the French Revolution. In the end, however, it did succeed in changing the social imaginary of French society and of the West.

While the notion of "a people" can be seen to have roots in earlier history (in a sense, the whole of Taylor's account shows the gradual emergence of both the institutions and the developing idea of "a people"), both American and French revolutions are threshold events in which the modern social imaginary embodies the notion of "a people" as the self-founding and self-governing agency of nations and communities in an integral form. This form continues to extend its influence as more and more groups assume agency and claim for themselves the status of "we, the people."[28]

Thus Western modernity is a social imaginary that embodies three major concepts. The economy is understood to be the form of human relations whereby acting for one's own purposes enmeshes with and generally furthers the purposes of others. There exists a public sphere governing the polity of the people. The people are self-constituted and self-governed. This reference to self, independent from any sacred or predetermined order of time and space, is characterized as secular. A shift occurred, which, Taylor observes, gradually relocated religion from its previous position as the central ground of society to a new more tentative location. Furthermore, from an ecological perspective (not Taylor's) the severing of the psychic connection to cosmos as a mediation of meaning and value paved the way for the virtual disregard of the natural world and the nature of humans as creatures of the earth and the physical universe. It could be argued, alternatively, that the emergence of natural law as a philosophical and scientific foundation replacing divine revelation with the rise of secularity could strengthen the ties of humans to the natural world. Whereas such a potential existed and does exist, the scholarship suggests that natural law did not have this effect, at least in any substantial way. In Western modernity no significant structure or set of practices required legitimation from principles considered inherent to the nature of the physical universe. Natural law itself carried various meanings and has diverse traditions. For Christians (who accepted it) it was understood as the law of God implanted in human consciousness and became a standard for moral decisions.[29] In the scientific world natural law referred to a deterministic ordering principle, which in the early stages, was associated with the Creator, but later came to be seen as the result of evolution based purely on chance. Another variance says that natural law is imposed on nature by the Creator; nature itself otherwise being disordered and chaotic.[30] In any case, natural law and with it the idea of nature itself was highly abstract and had little to do with the concrete natural world that was more and more to become the instrument of human progress.

Religion in the Social Imaginary of the West

As the key forms of the social imaginary changed in the West, so did the role of religion. Understanding the role that religion has come to play in the modern social imaginary is critical, if we are to claim that religious writing about ecology matters. Secularity is popularly understood as the condition whereby God is removed from public life. All categories of life can proceed without reference to a higher or supernatural power. Hence, religious faith and practice are often seen as a throwback to the premodern. Taylor's compelling argument about the relationship of religion to secularity allows for a more nuanced and, at the same time, more sure-footed approach to religion's power within modern societies. It shows with considerable specificity how religion can affect the social imaginary. This is important for ecotheologians who, although keenly aware of what was lost in the relation of humans and the earth in the transition to the modern world, know that no return is possible. A new scenario—a modification, perhaps even a radical one, to the Western social imaginary—is required.

Taylor concludes that the long cultural march to secularity that occurred in the modern West "removed one mode in which God was formerly present" but did not totally remove God's presence from the public space.[31] The present social imaginary is characterized by horizontal time, which means that all that is significant within the society is referenced in ordinary rather than higher or transcendent time. Actions for mutual benefit have always ordered human society, and in premodern times the origins were seen as created by God and the destiny was also with God (in Christian terms).[32] Such is no longer the case; God or some higher power is no longer an ontic necessity for any legitimate form or activity. Secular societies understand themselves as both founded and fulfilled in horizontal or profane time. This indicates for Taylor not the absence or end of religion and the relevance of God, but a new mode of God-presence in both personal and public life.[33] He deals with Christianity specifically because of its substantial presence in the history of the West.

As Taylor traces the rise of national (and other group) identities, he points to the complexity of forces, including religion, which enters into their formation. God can be present through devotion and a strong sense of God's will operating in one's personal life. Likewise, in public life, "God can figure strongly" in the formation of identities. "God's will can be very present to us in the design of things, in cosmos, state and personal life. God can seem the inescapable source for our power to impart order to our lives, both individually and socially."[34] While it may be wise, Taylor advises, to distinguish our political identity from any particular confessional stance, it will always be necessary to be attentive to changes in the interaction of religion and the state as long as the importance of religion persists as it does "virtually everywhere."[35] The compelling truth of Taylor's assertion has been demonstrated recently in both North America and Europe as immigrant religious groups vie for social and legal rights and recognition.

In light of Taylor's analysis, religious adherents must be aware of the larger social imaginary and the implications of its secularity if they are to negotiate religion's role within our present society. If they are to make a difference to the ecological crisis, they must be critically astute and confident with regard to their new role. This is not primarily to adapt to the social imaginary as some static reality, but to capitalize on the contingency of its present forms and its overall malleability in a process that will include at times negotiation and at times resistance.

ECOLOGY IN THE SOCIAL IMAGINARY OF THE WEST

The role of perceptions of nature within the Western social imaginary is particularly pertinent to the argument we are making. There are traditions within the West that can and have been revived in the ecotheology texts we discuss in the chapters ahead. However, these are, as most will admit, not the traditions that gained hegemony. Furthermore, many of the key developments in the West contributed substantially to many aspects of the ecological crises we experience today.

Nature in the Social Imaginary

In the social imaginary that Taylor describes, ecological consciousness has no substantial part in the long development of modern Western societies. As the notions of human freedom, individual rights, and the human control of history were strengthened, and the notions that confined human society to the destiny of God and cosmos were weakened, the concrete natural world—the soil, the water, the air—became simply instruments of Western progress. In his account of the social imaginary, Taylor does not deal with the ecological crisis or the role of nature, but he observes that nature as natural law is a normative ground for charters and constitutions based on inalienable human rights.[36] This is a nuance worth exploring as natural law understood in this fashion represents an abstraction. While maintaining some remnant of attachment to nature as a form of legitimacy for moral action, it is detached from actual natural phenomena. Furthermore, it comes more and more under attack as a basis for moral decision and practice in the Western social imaginary.

Bruno Latour extensively develops the latter point.[37] Latour differentiates Facts (capital F) from opinions, Nature (capital N) from nature in its concrete collectivity, and Science (capital S) from science as it is practiced in a constructed and tentative manner. For Latour the current impasse on the ecological front is not due to the absence of Nature in the Western imaginary, but to the mode of its presence—that is to say, Nature as an abstracted form rather than nature in its concreteness.[38] Latour's thematization is quite different from that of Taylor as well as from the many writers on ecology and religion who see nature as absent from modern society, but it does not contradict the main lines of that critique. Furthermore, his work offers insight into the way in which scientists as representative of Nature (even in its abstract form) can use their position to advocate on behalf of nature (in its concrete form).

CRITIQUES OF THE WEST FROM ECOLOGICAL PERSPECTIVES

Ecotheologians concerned with culture and society more than professional science find fault with the key forms within which the social imaginary, as Taylor describes it, took shape. The focus on the power of the human in history, whether through the economy, the public sphere, or the rights of individuals to participate in self-governance, is seen as detrimental to the nonhuman beings including planet earth itself. The excessive focus on the human is seen to have emerged from deep biblical notions about the human in relationship to the rest of creation. This focus is then exacerbated by the scientific and economic developments of the sixteenth and seventeenth centuries. A few of the accounts most influential for ecotheologians since the 1970s are those of Lynn White Jr., Thomas Berry, and Carolyn Merchant, although none of these authors was a professional theologian. These critiques will be presented and discussed in the chapters ahead.

Such accounts are consistent with Taylor's description of the social imaginary of Western modernity. They observe not so much the teachings and motivations of biblical religion or (in the case of Merchant) new scientific methods but the way in which these ideas and motivations came to be picked up and developed in actual practice. As Roderick Nash observes of White's critique, his "approach was pragmatic." The relevant question for White was not "what does Christianity mean?" but "what did it mean to a particular society at a given time and place?"[39] In other words, how did meaning and practice interact, a concern that is clearly also Taylor's. Furthermore, Taylor pays attention to the role that Christianity played in replacing some of the core beliefs about how humans saw themselves in relation to the rest of creation.

While Taylor gives credence to the complexity of the process of disembedding humans from the cosmos and the variety of factors that may have contributed, he pays particular attention to the role of Christianity in this process. Relying on the concept of the axial period proposed by Karl Jaspers, Taylor sees the roots of this disembedding in the last millennium BCE, when the biblical religion of the Hebrews, Confucianism, Buddhism, and the philosophy of Socrates appeared seemingly independently on different continents.[40] Taylor observes that these religions made possible the separation of religion from cosmic and social control that tied human activity to what were seen as established natural cycles; in Taylor's words:

> Perhaps most fundamental of all is the revisionary stance toward the human good in axial religions. More or less radically, they all call into question the received, seemingly unquestionable understandings of human flourishing, and hence inevitably also the structures of society and the features of the cosmos through which this flourishing was supposedly achieved.[41]

Whereas the process of disembedding continued with traces of a previous worldview, ties to the natural world that existed in pagan religions were in the process of being severed.

For Taylor, Christianity disenchants in two interrelated ways. It undermines the conviction that the reality of the given world established in Original time could not be changed. Christians aim to re-create the world as the Kingdom of God, "purging it of its connection to an enchanted cosmos. . . ." This building of the Kingdom of God relies on the belief that one is covenanted (called by God and invited to respond) to life in the Kingdom of God on earth not by virtue of birth but by choice. Originally hemmed in by the strong social controls of the established cosmopolitical order of King and Empire, Christianity did not reach full expression of these ideas until the seventeenth century. At that time the Protestant reformation gave force to the idea of a society "founded on covenant and, hence, as ultimately constituted by the decisions of free individuals," as discussed above.[42]

Taylor is not unique in this account of the effect of Christianity on the moral order of the West. What Taylor adds is a clarification of the process by which the social imaginary is shaped in the interplay of ideas and practices. The sense of the individual as an agent of society was first embraced only by elites (and mostly as an idea) in European society. Yet the idea influenced the social imaginary in mutual interaction with social and economic practice, and indeed may well have emerged from social and economic practice. Thus he argues, "Certain moral self-understandings are embedded in certain practices, which can mean both that they are promoted by the spread of these practices and that they shape the practices and help them get established. It is equally absurd to believe that the practices always come first, or to adopt the opposite view, that ideas somehow drive history."[43]

Taylor's account of the role of Christianity in the social imaginary of the modern West provides two important insights with regard to the ecological crisis. First, there is support for the persistent contention that Christianity played a critical role in Western attitudes and practices toward nature. Hence Christianity is complicit in the ecological crisis, because the disembedding of humans from the cosmic-based social and political order, to which Christian belief and practice contributed, served to disenchant and instrumentalize the natural world. Second, the social imaginary is a fluid and changing reality; new ideas and practices in mutual interaction in the sociopolitical world can have profound effects over time. All this means that while the current social imaginary of the modern West does not effectively include the ecological imagination, present efforts for ecological change are working. And it is precisely through a greening of the social imaginary that positive ecological change will happen. Greening in this context means much more than a decorative or trivial appearance of green, or a few adjustments to the existing systemic order. The depth to which the present social imaginary is not green only serves to indicate the extent of the challenge that lies ahead and the degree of intentionality that must be brought to bear to penetrate our taken-for-granted and convenient ways of thinking and acting.

ROOTS OF GRIEF AND HOPE—
A THEOLOGICAL INTERPRETATION

The dynamics of change that Taylor describes are influenced by the normative values held and expressed by members of society at any particular time. With regard to the modern West, it would be inconceivable to speak of change without recognizing the major influence of Christian concepts and values (as Taylor does). Because of the fluid way in which ideas and practices interact and either reassert or change the social imaginary, however, it is easy to see how nonintentional change or outcomes of change can emerge. The assumptions of the social imaginary are generally not clearly observed or articulated, most likely not even understood at any particular time. As a result there is the phenomenon that social scientists call "drift,"[44] which refers to the very gradual accumulation of activities or ways of thinking that build up incrementally. Drift can be seen to be a concomitant part of all change. One might conceive of the present ecological state of the world as a result of such drift. No one would argue that our ancestors set out to destroy the foundation of human existence. Yet an accumulation of activities has created just such a dangerous state of destruction.

Christianity interprets such drifts or malaise as indicative of the need for the human mediation of the divine presence in history. A core concern of the ecological texts considered here is the need for human but divinely inspired agency to heal the present ecological malaise.

Canadian theologian Bernard Lonergan describes certain kinds of persistent and particularly pernicious evil according to what he calls the longer cycle of decline.[45] The cycle is "longer" by comparison to short-term dysfunctionality caused by individual egotism or group biases that temporarily inhibit true progress. What Lonergan means by progress is not the ideology of unlimited progress that has colonized the social imaginary of the West and driven the socioeconomic system, but rather the ways in which the human good is increasingly incarnated in judgments and actions. The longer cycle of decline occurs when common sense refuses to acknowledge the long-term views and critical theoretic analyses within a community or society. Typically, insights deemed to be impractical or contrary to immediate interests are refused. Over time, valuable insights for the guidance of society are lost from one generation to another.

Furthermore, the longer cycle of decline is often characterized by the isolation of cultural forms from the social reality. The arts and academy are instrumentalized for the support of the status quo. Religion is relegated to a purely personal role. In all, the separation of theory and practice, of culture and society, result in a situation in which all that is intelligible is the balance of economic pressures and national powers.[46] For Lonergan, as for Christian theologians in general, this account of long-term dysfunctionality in society is value laden. Dysfunctionality is assumed to mean neglect of the practice of such values as charity and social justice, as well as a cumulative loss of human creativity and freedom. It is, in more traditional language, a sinful condition. The longer cycle of decline calls for intentional and authentic human intervention. Ultimately, in Christian terms, authenticity requires a difficult and challenging response to freely given divine love.

For our purposes, Lonergan's sense of the longer cycle of decline alerts us to a distinctly religious interpretation in the texts of Christian ecotheologians. This is not a negation of the largely secular account presented above. Indeed, we are claiming that such secular accounts are critical in helping establish the concreteness of hope. There is no actual separation of Christianity and secularity. Religious believers inhabit and act in both religious and secular contexts. The stakes for change are intensified, however, in a Christian account. They are the stuff of salvation. This is the case that the ecotheological texts make. Thomas Berry, one of the primary thinkers influencing ecotheology, claims: "It is increasingly clear that none of the children, nor any living being on this continent or throughout the entire planet has any integral future except in alliance with every other being that finds its home here."[47] This will require a difficult negotiation, complicated by the peculiar location of religion in the social imaginary of the West.

Christianity claims to make a concrete difference in the world. The salvation story of the Christian tradition rests on change in human history, in the attitudes and practices by which its adherents live. While religion occupies quite a different place vis-à-vis the social imaginary today than it did in the premodern era, it continues to be a source of meaning and values to many. It continues to function, even if in less defined ways, in the dynamic formation of the social imaginary. Thus the ideas and practices that Christians and Christian institutions promulgate are a significant set of tools for bringing about concrete change. The religious motivations for engaged activity in the world, especially in resisting the pernicious tendencies to decline, can provide an important impetus for positive change.

In the chapters ahead, we explore the conditions by which Christian texts can create positive change toward a viable and worthy future. In the words of Bill McKibben, already noted in the introduction, "Real hope implies real willingness to change."[48] The challenge is to bring this sense of willingness into the concreteness of life, where change happens in the social imaginary, at a day-to-day pace often too incremental to observe.

HOPE AND VISION INCARNATE

The present social imaginary in the West has little in its repertory that is ecologically sane. The community of texts we examine in this book participate in the process by which this repertory can be extended and changed to include ecological values and practices. For the creation of such a value-laden repertory, Taylor calls for "a complex, many-leveled struggle, intellectual, spiritual, and political in which the debates in the public arena interlink with those in a host of institutional settings."[49] These debates must reflect the theoretic reflections, the lived practices, and the demands for human authenticity that are relevant to the debate. They must be set in the wider context no less than "the shape of human life and its relation to the cosmos." "But," Taylor adds, "to engage effectively in this many-faceted debate, one has to see what is great in the culture of modernity, as well as what is shallow or dangerous. As Pascal said about human beings, modernity is characterized by *grandeur* as well as by *misère*. Only a view that embraces both can give us the undistorted insight into our era that we need to rise to its greatest challenge."[50]

It is clear from Taylor's reflection on the social imaginary that visionary ideas are never perfectly incarnated in practice. Utopias, for example, have been imagined and described; they have never actually existed. What is equally clear, however, is that our only hope lies in the incarnation of the best visions for a better future. That future will be imperfect. We can confidently predict that we will never be perfect ecological citizens of the earth. Nor can we return to a premodern existence. The starting conditions for the incarnation of a new vision, at least as far as the West is concerned, are set by the contemporary social imaginary with both its strengths and weaknesses. Our future will be negotiated between what presently exists and what we are committed to creating. As in all negotiations, success is not measured by the full or immediate achievement of the ideal situation, but by the process of engagement.

2

The Emergence of Ecotheology

This chapter examines two contexts in which Christian ecotheological texts emerged—the sociopolitical conversation that heightened ecological awareness in the mid-twentieth century, and the theological conversation that eventually legitimized ecotheology in the academy. These contexts are substantially different. This is not to say that all awareness of ecology and even of a threatening crisis began in the 1960s. There are traditions such as the conservation movement, environmentalism, natural history, and transcendentalism that were operative and sometimes made important incursions into the dominant culture.[1] In the late nineteenth and early twentieth centuries, George Perkins Marsh, Gifford Pinchot, and John Muir (more correctly a preservationist) were instrumental in giving teeth to the conservation movement in the United States resulting in the establishment of forest reserves and national parks.[2] Canadian conservationists were also part of this movement and first met to discuss the same issues in Montreal in 1880.[3] Similar movements were occurring in Europe. Eighteenth-century tourism in Canada emphasized the pristine natural environment.[4]

Some of those who came to prominence within the ecological movement of the late twentieth century drew on these previous resources. Thomas Berry, who was an important bridge figure between the Christian ecotheologians presented below and the broader cultural history, is preeminent among those who incorporated these historical movements into the construction and presentation of his own vision.[5] There is no doubt that many of these existing movements contributed to the invigoration of ecological awareness and activity in the period we discuss below. However, we choose to focus more closely on this much shorter and more recent history of the incursion of new ideas and activity that give space and energy to the emergence of ecotheology in the mid- to late twentieth century.[6] This closer look demonstrates more clearly how ecotheology, by participating in such a movement for change, shares also in the dynamics of changing a social imaginary in a desired direction.

While ecotheology brought a different perspective and challenged a whole new identifiable public—that of Christian theologians and adherents—the imagination of a new ecologically responsible world was becoming front and center in the 1960s and 1970s. The Christian texts that are the subject of this book entered a public conversation. There were already vibrant social movements, many of which included or had predominant ecological agendas. Rachel Carson had published *Silent Spring* in 1962 decrying the use of toxic chemicals as pesticides (especially DDT); Fritz Schumacher, Barry Commoner, and others were already bringing ecological consciousness into leftist political movements. Campaigns against testing of the atomic bomb had ecological as well as peace agendas. Previously existing ecological and nature foundations were taking a more radical turn. Greenpeace was established.[7] The social and ecological movements were incursions into the existing social imaginary. As such, they provided an opening within which Christian theologians could begin reflection and conversation on the relevance of Christian belief and practice to ecology. Furthermore, these movements actually fueled the conversation in academic Christian settings. Christian theologians found within the society at large (1) a language from science, social and civil rights, and political activism, and (2) a challenge to the churches to recast their theology and practices, both action and meaning, in response to the modern world. For some, this included the ecological crisis.

While the stage was set for Christian theologians to take their places within this new space, however, the emergence of ecotheology within the academy had all the characteristics and challenges of a scientific intellectual movement. We examine this in detail in the second part of this chapter.

SOCIOPOLITICAL CONTEXT OF THE 1960S AND 1970S

No matter how one might judge the rhetoric and practices of the youth who came of age in the 1960s, their political activism for a more peaceful and just world set the stage in the public sphere where imagining a new kind of life together and a new sense of responsibility to the wider and future world could be explored. On the ecological side, their activism included experiments with "back to the land" movements, simple communal living, and resistance and repudiation of the perceived artificiality and hypocrisy of the inherited society. Those in North America and Europe who marched against the U.S. War in Vietnam and the existence of nuclear armaments couched their actions in rhetoric about the planet, even if the extent of the ecological crisis was not yet widely known.

Space exploration has been called the cultural icon of the 1960s and among its many effects was the creation of a picture of the entire planet.[8] Although later contested in their meanings and political significance, the photographs of *Earthrise* (1968) and of *The Whole Earth* (1972) and the heady optimism of the moon walk of 1968 raised the expectations of what was possible in terms of breaking human boundaries. At the same time these images increased attitudes of solidarity among many people toward the protection of one human family and one home, Earth.[9] Not only was the Earth impressive in its beauty, but it also seemed fragile and destructible.

Key Groups and Personalities

Rex Weyler's account of the founding of Greenpeace illustrates how the roots of present ecological organizations are clearly tied to anti-nuclear and peace movements of the 1960s.[10] The "Don't Make a Wave" campaign to stop the further testing of nuclear weapons by the United States emerged after the 1965 test under Amchitka Island. Just prior to the second test in October 1969, 6,000 protesters assembled on the border between Washington State and British Columbia, Canada (one of many similar protests in North America and Europe). The protesters included activists, mostly young people from universities and secondary schools, many of whom formed the core of the emerging new environmental movements.[11]

In her account of the ecology movement in the twentieth century, Anna Bramwell chronicles the transition of ecology as a movement in the 1960s and 1970s as a process where the perceptions of ecologists and nature preservationists (as conservative, elite, and nationalist, as they generally had been classified up to that time) became more populist and left leaning, even radical.[12] The well-known Nature Conservancy Council was the subject of such a change. Founded in England in 1949 with pressure from the Royal Society for Nature Conservation, this agency was dedicated to the creation of nature preserves and ecological research. In Bramwell's words, "The Nature Conservancy Council experienced an important internal revolution in the late 1960s, designed to change it, according to the radicals, from a powerless, underfunded group of elderly conservationists to a politicized, dynamic pressure group, worried about mounting threats to the biosphere."[13] According to Bramwell's analysis, quite a number of new concerns emerging in the 1960s and early 1970s conspired to influence a new ecological movement. Especially in North America and Western Europe the optimism that accompanied the rebuilding of Europe after World War II and the growing consumerist and technological society in the United States began to come under severe questioning: Was bigger always better? Did all these new technologies work for the poorer nations? Did they indeed deliver on the promises of a better quality of life for anyone? A growing awareness of the inability of the richer nations to solve the poverty of the world, to deal with increasing populations, and counteract the legacy of colonialism contributed to the radical oppositions especially of the young.

Rachel Carson and Silent Spring

Bramwell notes some key iconic events and personalities of the ecology movement. A brief examination of these reveals a gathering sense of solidarity around ecological concerns. In 1962, Rachel Carson's *Silent Spring* brought the issue of chemical pollution to the fore. Many of the claims made by Carson in *Silent Spring* can be found in scientific journals and newspapers predating its publication, but Carson was able to compile the data and present it in readable fashion to the general public. In effect, she was responsible for increasing dramatically the scope of the discussion about ecology by appealing to a general audience that did not have access to the scientific terminology in which previous reports were usually cast. Many of her arguments and exhortations appealed to young radicals in North America who were appalled at the materialist and wasteful nature of modern capitalism. In one such exhortation she exclaimed:

> It is also an era dominated by industry, in which the right to make a dollar at whatever cost is seldom challenged. When the public protests, confronted with some obvious evidence of damaging results of pesticide applications, it is fed little tranquilizing pills of half-truth. We urgently need an end to these false assurances, to the sugar coating of unpalatable facts. It is the public that is being asked to assume the risks that the insect controllers calculate.[14]

The rhetoric and the cause echoed the voices of disillusionment with the establishment commonly expressed by the youth of the day.

It has since been pointed out that Carson was not advocating the complete abandonment of scientific technology in controlling the natural world. As Lisa Sideris demonstrates from Carson's own reflections on *Silent Spring*, Carson's primary concern was the poisoning of the natural world, not of humans alone. "In the end, the book was both about poison and about life; it was, and is, about humans and nature. *Silent Spring* demonstrated that we cannot talk about one of these things without talking about the other. But first and foremost, Carson believed, we must *talk* about them."[15]

But it was not only the "young radicals" who were moved and in some cases provoked to action. A letter to Carson represents the sentiment of another segment of the public, those city dwellers nostalgic for the "nature" they missed. One excerpt from a writer, identifying as such, claims, "You are a poet not only because you use words so well, but because by describing non-human life without pathetic fallacies you make us readers understand our place on earth so much better."[16] Serious conservationists in responsible positions concurred. "Nowhere have I read a clearer exposition of the subtle and cumulative effects of certain kinds of poisons," wrote a professional biologist, Hampton Cross, in a book review.[17] Nobel Prize winner Herman Muller described Carson's "most important service" as enlightening the public about the "complexity and interrelatedness of the web of life in which we have our being."[18]

Not all reactions were positive; Carson's *Silent Spring* provoked the ire and backlash of many, especially the petrochemical and pesticide industries and the U.S. Department of Agriculture. The National Agricultural Chemicals Association treated the issue as one of public image and staged a well-funded campaign to counteract the perceived negative effects of Carson's book. The details of such campaigns have been well documented.[19] Public debate and controversy succeeded in bringing this ecological issue to the fore and one of the more immediate signs of the success of *Silent Spring* was legislation by the U.S. government limiting the use of pesticides, especially of DDT, in 1973. Rachel Carson's *Silent Spring* engaged a number of social sectors in position to hear what she had to say; it also brought these sectors into dialogue and engaged the general public on an ecological issue. Hitherto fragmented pockets of awareness and engagement were united, if not in agreement, at least in a common concern. With reference to the critical role of good communication in the success of new ideas, J. R.

McNeill, among others, considers Carson "the single most important catalyst for environmentalism."[20]

Fritz Schumacher

As Bramwell notes, the politicization of long-standing nature conservation societies both contributed to and are evidence of a growing public engagement with ecological issues in the 1960s and 1970s.[21] In 1971, Fritz Schumacher, author of *Small is Beautiful* and founder of the Intermediate Technology Movement, took over leadership of the Soil Association (founded 1945) of the United Kingdom. Barry Commoner, an American environmentalist, who saw the multinational corporations as primarily responsible for the proliferation of technologies and the growth of consumerism, was soon appointed vice-chairman.

Schumacher was an economist who earned his stripes working in important advisory positions on the British Control Commission in Germany (1946–1950) and then on the National Coal Board in England for 20 years. In the early 1950s he visited Burma and became impressed with Buddhist economics. Already disenchanted with the economic systems of the West and their failure to aid the poorer nations (as well as the poor in the richer world), Schumacher realized two important ideas that occupied the rest of his life and influenced many. He was convinced (1) that economic practice is driven by a meta-economics, a set of ideas about materialism and progress that undergirds all economic policy and activity; therefore a change of philosophy and outlook was needed, and (2) that "small is beautiful," that is, that intermediate technologies were necessary for economic development. The practice of transporting large Western advanced technologies into the so-called developing nations was not only failing but also devastating the societies they were intended to serve.

Schumacher was not only sought out by governments within the majority world (Tanzania, 1968, for example), but also acquired a large following in Europe and North America. When he published *Small is Beautiful* (1973) on the heels of the UN Conference on the Environment (Stockholm, 1972), the book was consumed almost instantaneously by audiences as diverse as the eco-warriors of California, the British Monarchy (Schumacher was recognized on the Queen's Honours List, 1974), the White House (President Jimmy Carter would later insist on a photograph holding the book, 1977), and governments of the majority world. As was the case with *Silent Spring*, Schumacher's *Small is Beautiful* gave a focus to existing ecological concerns and created further concern. The readable combination of economic theory, its philosophical premises, and its practical and hopeful solutions increased the scope of the ecological movement as it also strengthened ties between social and environmental issues. Even more so than the work of Rachel Carson, Schumacher's work had deep religious personal and ideological roots both in Buddhist economics and Roman Catholic social teaching. He had direct influence on the earliest religious ecologists, such as Thomas Berry and John Cobb Jr., supplying them with ideas, inspiration, and conference opportunities for discussion and publication.

Barry Commoner

Barry Commoner was a biologist who specialized in the study of the effects of chemicals such as chlorofluorocarbons on the natural world. In the late 1950s he emerged on the public scene with an outspoken protest against atmospheric testing of nuclear weapons. In the 1960s, he became an active environmentalist concentrating primarily on the petrochemical industry and its deleterious effects especially on the ozone layer. He also engaged discussion on a wide range of concerns including pollution and population. In *Science and Survival* (1967), Commoner warned of the long-range effects of the rapid growth in technology especially on the environment, and he outlined the responsibility of scientists to be accountable to the public. *Time* (February 2, 1970) had already dubbed Commoner "the Paul Revere of ecology" because of his outspoken warnings of an increasing environmental crisis. Commoner's most popular book was *The Closing Circle*,[22] in which he developed the notion that humans had broken the circle of life. He gave concrete scientific explanations of the human interdependency with the earth and the threat posed by human activities such as nuclear bomb testing, industrial pollution, overpopulation, increased technological power, and excessive use of fertilizers and pesticides. It is noteworthy that Commoner's starting point for *The Closing Circle* is an overview of Earth Day in 1970, including snippets of speeches and comments from across the United States.[23] Activism was strong, he concluded, but the diverse and even contradictory advice posed at the Earth Day events reflected a lack of scientific understanding of the environmental crisis.

Commoner wrote scientific explanations in a popular and compelling manner. His "First Law of Ecology—Everything is Connected to Everything Else" remains a core assumption for all ecologists and environmentalists. Furthermore, Commoner was early among environmentalists in pointing to the necessity for alternate technologies, especially in production processes. Like Carson and Schumacher, he did not stop at giving explanations. He called for concrete and specific policy and action changes on a large scale. When the first UN Conference on the Human Environment was held in Stockholm, Sweden, in 1972, Commoner spoke at a "counterconference" organized mostly by women who were protesting the under-representation of women at the official conference. Commoner was an early voice for inclusion of issues such as poverty and unsound production of goods in the discussion of the ecological crisis. He felt that such issues were not adequately addressed at the UN conference.[24] As was the case with Carson and Schumacher, Commoner brought focus, deeper understanding, and powerful communicative skill to the environmental movement that was emerging in the wider public domain. As he himself observed of the First Earth Day in 1970, the movement had a lot of energy and enthusiasm but was suffering from its own diversity and lack of clear scientific understanding of the environmental dangers it protested.[25] The need for leadership and communication was characteristic of the young activists of the 1960s and early 1970s. Characterized for the most part by a countercultural mentality and a desire for individual freedom, the hippie "back to nature" movement was not an organized movement, nor were the hippies politically effective. However, they did listen and respond to the environmental critiques of Carson, Schumacher, and Commoner.[26]

EARLY ROLE OF RELIGION

Religions (certainly organized traditional religions) do not seem to play any significant and identifiable leadership role in the emergence of the ecology movement of the 1960s and 1970s, but they did participate. Certainly, there was a spiritual element to many of the countercultural and back-to-the-land movements. A recognizable group of neo-pagans emerged. The neo-pagans left traditional institutional religion and opted for a version of pre-Christian earth-centered religion that they claimed was more ecologically responsible, more peaceful, and more open to equality. While the neo-paganism that developed from the mid-twentieth century on has various identifiable groups and diverse beliefs and practices, it was the attraction of nature and the unity of humans and nature that appealed to the first of these young counterculturalists.[27] Many were peace and environmental activists. The best known and perhaps also the icon among the neo-pagans was the author and activist, Starhawk. Starhawk was a Vietnam War protester and peace activist as well as a neo-pagan practitioner (Wiccan variety) and environmental advocate. She continues to write, teach, and protest on behalf of peace and ecology today.[28]

More traditional religious groups were present *en masse* for anti-nuclear demonstrations that included ecological concerns. For example, some of the first movers in the organization later called Greenpeace self-identified as Quakers and Unitarians.[29] The Society of Friends' Quaker Earthcare Witness acknowledges its roots in the peace and environmental movements of the 1960s.[30] Joseph Sittler, one of the very earliest Christian theologians to formally respond to the environmental crisis, began his sermon "The Care of the Earth" in 1963 by noting the contemporary human anxiety presented eloquently in a poem by Richard Wilbur. He commented with obvious reference to the threat of nuclear disaster, "The substance [of the anxiety] is this: annihilating power is in nervous and passionate hands. The stuff is really there to incinerate the earth—and the certainty that it will not be used is not there."[31]

Bernard Lonergan describes the function of theology as mediating between a religious tradition and the cultural context.[32] Such was the acknowledged case for Christian theologians who sought to recover and recast the tradition in various ways in the light of the crises (ecological, poverty, oppression, nuclear war, and others) as they came into heightened awareness in the 1960s and 1970s. Pierre Teilhard de Chardin had addressed the new developments in biological sciences in a new Christian understanding of evolution.[33] Political and liberation theologies were emerging in an effort to relate Christian faith to an increasingly secular world in the North and an increasingly poor and oppressed world in large regions of the South.[34] On the one hand, the cultural crisis demanded a theological response; on the other, movements within the academy itself, such as the renewal of theology and the opening of theology to other disciplines, in particular physical and social sciences, provided an intellectual environment for the emergence of a theology of ecology.

As noted at the beginning of this chapter, Christian ecological texts can be understood differently but in complementary ways by considering two of the significant contexts out of which they were constructed. We have discussed the sociopolitical context of the mid-twentieth century, which to a large extent was already aware of the ecological crisis and often active about it. We now turn to theological academies and Christian seminaries where a spirit of reform of theology and Christian practices challenged the status quo. Changing a social imaginary is an intricate and certainly dialogical process in which ideas and practices spark the imagination of groups and individuals who carry forward some version, at least, of these ideas and practices. This is partly the result of the readiness of some groups and individuals or the particular kind of fit that exists for what is new, and partly the result of the power of the emerging ideas and practices themselves. The development of a green consciousness in theological circles and institutions is a case study of an emerging, if localized, change in a social imaginary.

ECOLOGY AS A SCIENTIFIC INTELLECTUAL MOVEMENT

The structure developed by Scott Frickel and Neil Gross in "A General Theory of Scientific Intellectual Movements"[35] provides a useful and critical framework for description and analysis of the development of texts within institutional contexts. Especially useful is attention to politics, power, and resistance, and the authors' grasp of the particularities of the academic world (in which the majority of our texts were developed).[36]

Scientific Intellectual Movements (SIMs)

The most abbreviated definition of scientific/intellectual movements (SIMs) is this: "SIMs are collective efforts to pursue research programs or projects for thought in the face of resistance from others in the scientific or intellectual community."[37] Two elements stand out: "collective efforts" and "in the face of resistance from others."

Ecotheological texts are in some sense a collective effort—that is, they are joined together by intent—a "correcting of long-held wrong doctrines, the call for changes in action, and reflection on all sorts of issues in light of the crisis we face,"[38]—and by a "knowledge core [towards which] participants are consciously oriented, regardless of their understanding of it."[39] The knowledge core comes primarily from attention to eco-degradation and to secular literature and movements described in the first part of this chapter.

The conversation was not a collective effort in the sense that the texts were held-in-common and regularly referenced each other. In fact, comprehensive cross-referencing of authors even when interests intersect is rare. These authors who are joined together by intent do not reference other high-status contributors to the field, although they may have interacted informally at scholarly conferences. In large measure there appear to be fairly circumscribed circles of interaction—for example, circles emanating from Teilhard de Chardin, from process theology, from women's studies, or from Niebuhrian ethics.

More in common is the fact that these are texts developed in the face of resistance from others. Frickel and Gross argue "a movement is a SIM by our definition only if, at the time of its emergence, it significantly challenges received wisdom or dominant ways of approaching some problem or issue and thus encounters resistance."[40] Of the response to Joseph Sittler's address to the World Council of Churches Assembly in New Delhi in 1961, where Sittler claimed, "the sphere of grace and redemption can be no smaller than the sphere of creation itself,"[41] H. Paul Santmire said, "it was mainly one of polite indifference, along with some shocked resistance on the part of representative from the then reigning theological guilds in Europe."[42]

For some, the resistance was ideological. For example, people who interpret the Bible literally will likely resist a reread of the scriptures with an earth hermeneutic. Or people who are content with status quo relations between women and men, and between rich and poor nations are likely to resist the interpretations—and their consequences—of an author like Rosemary Radford Ruether. She writes, in *Women Healing Earth*: "This volume of essays by women in Latin America, Asia, and Africa on religion, ecology, and feminism presents an effort at cross-cultural communication and solidarity between women in the 'First World' and women in those countries that are struggling against the effects of Western colonization. . . . By viewing the ruling classes of my country from the underside, its evils and lies are revealed and put in the context of a larger reality and call for justice."[43]

Resistance within the Academy

Within the academy, resistance may take a quite different form. John Cobb Jr. describes it this way:

> Indeed, the academy is peculiarly resistant. Greening involves seeing things in their interconnectedness. The academy is structured around the division of subject matters and into departments. Greening involves a sense of relatedness or unity that is deeply subjective and then also implies kinship in what is felt. The methodology of the academy involves the strict control, if not the elimination, of the student's feelings and the neglect or denial of the feelings of what is studied, whether that is human or not human. As long as theology is a university discipline seeking respectability in that context, it is hard to see how it can be greened very far.[44]

Frickel and Gross put it another way: "Precisely because the intellectual practices recommended by SIMs are contentious, SIMs are inherently political.... Every program for intellectual change involves a desire to alter the configuration of social positions within or across intellectual fields in which power, attention, and other scarce resources are unequally distributed."[45] Of course, it is evident that the ecotheologians of our texts are politically oriented. To a person they care passionately that the political world be reshaped in harmony with the earth of which humans are a part. To a person they understand that political action is needed, and one can expect that with their individual and collective years of experience, they have learned that political change in an eco-friendly direction has never adequately succeeded. As Cobb notes, "But if we understand the real problems of the real world in any depth, we will see that specifically human problems are not really separable from the larger reality in which human beings are embedded."[46]

The political reality of a SIM is that it is shaped by the culture in which it comes into existence. As Frickel and Gross explain, "It is characteristic of SIMs that they are influenced by cultural and political pressures. Impetus for change can come from the larger social environment, which includes the state, religion, industry, education systems, and social movements."[47] However, the dynamics of academic institutions themselves play a role. Anyone who has worked in an academic institution understands, for example, the near impossibility of establishing a new course as a degree requirement.

It is because of the inherently political nature of academic institutions that the SIM that ecotheological texts represent emerged, in large part, because of the work of "high-status intellectual actors . . . [who] become cognizant of important theoretical shifts occurring in other intellectual fields, wish to incorporate those shifts into their own work, and become dissatisfied that this is not taking place . . . Whatever the processes involved, those who developed the initial theoretical or research program of a SIM, or who join its ranks shortly thereafter, did so, out of personal conviction and in spite of professional risks."[48] The interaction of events and texts in the wider community (e.g., Rachel Carson's *Silent Spring*) influenced some ecotheologians to break from an accepted academic path, and so risk losing power—credibility, students, publications, and so on. "Nevertheless, a SIM is ultimately dependent on the contributions of its intellectual leaders, who articulate its program and do the intellectual or scientific work that comes to be seen as the hallmark of the movement."[49] In all this interplay of resistance, power, scarce resources, and publics, high-status intellectuals produce texts (as well as teach when opportunity arises, which is not always possible within a particular academic program) and, in effect, profoundly reshape the social imaginary of their community of discourse. A successful SIM is, by definition, a reshaping of a particular social imaginary. As SIMs are in large measure fueled by texts that are cutting-edge, the link between a particular social imaginary and texts is clear, solid, and demonstrable.

Mobilizing Structures

Frickel and Gross take into account the role of network-based mobilizing structures, an example of which is an institute, conference, or a group with ongoing status in a scholarly organization. "For a SIM to be this successful, it needs access to organizational resources, or what scholars of social moments called mobilizing structures. As a great many sociologists of science have argued, science and academia are largely network-based, at various levels."[50] "Several types of micro-mobilization contexts for SIMs exist. For example, conference and symposia offer space for the incubation of new ideas, findings, or problems among like-minded but geographically separated thinkers."[51] Since the early 1970s, these have taken a variety of forms—conferences, cooperative book projects, ongoing study groups, and so forth. Holy Cross Centre at Port Burwell, Ontario, Canada, was an example of one such micro-mobilization context. There, for 22 consecutive summers, Thomas Berry and a widely diverse group of leaders in the field met for shorter or more extended periods with students, scholars, and interested lay people. The interplay between senior and junior scholars was particularly important.

Another example of a micro-mobilization context is FORE (Forum on Religion and Ecology) organized by Mary Evelyn Tucker and John Grim, now of Yale University. Its website describes FORE as "the largest international multireligious project of its kind." FORE facilitates conversations between religions and other disciplines (e.g., science, ethics, economics, education, public policy, gender) "in seeking comprehensive solutions to both global and local environmental problems."[52]

Finally, it should be noted that the American Academy of Religion (AAR) has had, since 1991, a group on Religion and Ecology. Its website provides an excellent history of the project.[53] Perhaps most importantly, it also lists all conference themes and major presentations from 1991 to the present.[54] The group has had notable success in recruiting new members and keeping existing members, thus extending the life of the SIM over several generations, and making links with a variety of disciplines.

FORMING A GREEN PROFESSIONAL IDENTITY

As Frickel and Gross maintain, "A professional identity is not guaranteed by the formation of learned societies and the production of extensive propaganda, necessary as these are. It also requires the recruitment of followers and students and more especially the creation of satisfactory career structures."[55] The example we now describe—the program in ecology and theology in the Faculty of Theology at the University of St. Michael's College, Toronto, Canada—is both conversation and a result of conversation. That is, as currently structured, it embodies a many decades-long immersion in broader public and theological/ethical developments and it carries the convictions that arise into an institutional, ongoing framework, which in turn guarantees that the conversation will continue.

In broad outline, the result of all these embodied conversations is that students in any graduate degree program (MRE, MDiv, MTS, ThM, PhD) can choose to do a specialization in Theology and Ecology. From the perspective of satisfactory career structures, this is a brilliant plan. A student who has specialized in Theology and Ecology has a recognizable professional degree on the basis of which to be hired. The specialization may or may not be of primary interest to the hiring entity, but the graduate is able to both train in and pursue interests in this specialization.

Our example demonstrates the way in which a SIM took root in one Faculty of Theology. While Frickel and Gross note the function of scarce resources, resistance, publics, and power, this example demonstrates how power can become cooperative with the right (and skilled) leadership. Stephen Dunn, who taught Ethics in the Faculty of Theology, University of St. Michael's College in Toronto, describes the Canadian experiment that he calls "Three Shades of Green."[56] Dunn had been a confrere for decades, personally and academically, of Thomas Berry. Over the years his own teaching of ethics and leadership in the Roman Catholic religious community in Toronto and southwestern Ontario had shifted toward eco-ethics. The role of texts was crucial. Berry's own work and support for wider reading in the field were key elements in the "Three Shades of Green." Dunn found encouragement from students at a time when faculty colleagues seemed to "see little relevance to theology in questions of acid rain or desertification." He wrote: "Even though theology students at the graduate level tend not to be so very young anymore, there was a genuine sense of generational difference—the planet at peril meant something profound to them. . . . From this positive experience, a conviction began to grow in me that the students were key to the changes in the educational structures the faculty could sponsor."[57]

In response to pressure from students and a perceived sense of readiness because of the sensed urgency of the ecological crisis, the faculty decided to "make a contribution to the healing of the Earth in all its life systems, and express its concern for the ecological crisis of the planet."[58] The mechanism for this was a new institute for theology and ecology. What Dunn faced in making this new institute an academic reality was daunting. There were resource constraints, no-hiring policies, and increased workloads and crowded schedules for faculty. All this took its toll in the time and creative energy needed for new initiatives. Yet Dunn argued his case to colleagues that the moment had arrived for new educational leadership.[59]

The Faculty of Theology in which Dunn was teaching is part of the Toronto School of Theology,

> which is arguably the most operationally functional ecumenical cluster anywhere. This provides a widely diversified and professionally impressive network of theology professors and courses. This became the primary element of an "'eco-systems'" approach. The problem then could be seen as different from a lack of resources. Rather, the challenge was to further activate the already present system of resources. The existing network required a new element to make it responsive to the challenge.[60]

Dunn convinced his colleagues that the new element was ecological studies. The strategy Dunn proposed was what he called "Three Shades of Green."

Deep Green involves core courses that deal directly with eco-theology (e.g., Christian Ethicists and Ecology, or Feminist Theology and Ecofeminism) and cross-disciplinary courses dealing with ecological studies. Dunn found great support in the university for collaboration and student cross-registration in courses as diverse as Global Ecological Issues from a Community Education Perspective, and Human Environment Theory I: Great Lakes Issues, Attitudes and Evidence.

Intermediate Green is a category of courses that "pursue, without necessarily making specific reference to, the ecological crisis, what might be called essential elements of new paradigm thought. . . . Ultimately they shape a context for how theology will relate to the new science and the new cosmology."[61] It is critical that people teaching these courses welcome students who are pursuing ecological interests. In this category Dunn identified courses in areas such as Process Theology, Religion and Science, and Theology and Native Peoples.

Light Green describes the rest of the courses of the theological spectrum—courses that make no specific reference to ecological issues, where professors would not themselves be comfortable making ecological connections. However, when professors are open to it, "students can, for example, isolate certain issues in Church history . . . , such as a deeper understanding of St. Francis, or Patristic openness to Creation, or, in a course on Christology, exploring with greater intensity the implications of cosmic Christology."[62]

The mechanics of how this program led to a certificate of specialization in each of the graduate programs is not important to this illustration of the actual local development of a SIM in a specific academic setting. What is important is that "a SIM is ultimately dependent on the contributions of its intellectual leaders, who articulate its program and do the intellectual or scientific work that comes to be seen as the hallmark of the movement."[63] In all this interplay of resistance, power, scarce resources, and publics, there is a profound reshaping of the social imaginary of a community of discourse, fueled both by high-status leadership and texts that are cutting-edge.[64]

FROM SIM TO SOCIAL MOVEMENT

Frickel and Gross conclude their study with a comparison of SIMs and social movements. "At their extreme, social and scientific/intellectual changes can become revolutionary, but in most cases, the SIMs are not likely to reverberate in the lives of most people, and . . . are obviously far more constrained than the civil rights and peace movements."[65] There are two issues that need to be addressed to further our contention that these texts, cumulatively, are a practice of hope that may influence and change the wider social imaginary. First, can ecotheologies pass forcefully into the general culture? Second, is the leadership exhibited by ecotheolgians collectively the kind of leadership that is likely to shape a social movement?

The texts and their authors with which we are concerned are well described using the categories developed by Frickel's and Gross's general theory of scientific/intellectual movements. "Like social movements, SIMs represent contentious challenges to normative practices and institutions and, as such, are inherently political. Requiring ongoing coordination at various levels of organizations, SIMs are episodic creatures that eventually and inevitably disappear, either through failure and disintegration, or through success and institutional stabilization."[66]

Where, at this point in time, is the SIM identified with ecotheology texts? Clearly, it has not failed or disintegrated. It has achieved a level of success and institutional stabilization. It has achieved legitimacy in the academy and in the mind of publics who read these texts. At every level of intellectual inquiry and scholarship, it is considered legitimate to factor in or teach about ecotheology. This is so whether one's academy is a seminary (and thus, e.g., in Bible, theology, ethics, religious education, pastoral care, preaching, and worship—or a course on ecotheology itself) or a religious studies department (and thus, e.g., in world religions, sociology of religion, contemporary religious movements—or a course on ecology and women's poverty). At this level, an ecological thrust in any course could be as easily defended as teaching any course from, for example, a feminist or postmodern perspective.

But have the texts and authors we are studying achieved the status of a social movement? "SIMs are rarely contentious in the same way that political activism is contentious. . . . Successful SIM tactics tend to involve mundane actions directed at contentious ends, and we would expect to find a far more subtle blurring of boundaries between collective action that is normative and that which is quietly transformative."[67] Thus, it may be difficult to assess the transformative effect of ecotheogical texts.

In part, this is so because ecotheologians are not generally revolutionary charismatic leaders that Frickel and Gross point to as typical of historically important social movements—Lenin, Mao, or Castro (to cite some of their examples), or Martin Luther King Jr., Rachel Carson, or Desmond Tutu (to choose a few of our own). But this is not the whole picture. We might, following Frickel's and Gross's argument, expect transformation from SIM to social movement for quite another reason, namely what is at stake for SIM participants. Typically SIM participants might, especially in the early history of a movement, feel they are risking professional prestige, but few "risk their lives."[68] Still, without exception ecotheological texts have a level of urgency and passion that shows that their authors are thoroughly convinced that the issues they write about are life and death issues for the human race, even for the planet as, for example, when John Cobb Jr. asked, in the title of his 1972 book, *Is It Too Late?* This was no merely academic question.

WHEN "NOTHING IS ECOLOGICALLY INNOCENT"

There is another dimension to the inquiry whether the SIM of ecotheology has achieved the status of a social movement and to what extent it is transformative. The question needs to be answered in a current social context that is different from that of the 1960s, 1970s, and 1980s.[69] How have ecotheology's publics changed and developed? George Myerson argues that the way in which society sees the urgency of ecological issues has changed dramatically in the last decade. Myerson compares Freud's insights on the unconscious in *The Psychopathology of Everyday Life* to what he calls *The Ecopathology of Everyday Life*, and argues that there is a growing realization that "nothing is ecologically innocent."[70]

> Just when you thought you had tamed Psychopathology, when it seemed safe to venture out among the trivialities again, here comes its even more threatening cousin, Ecopathology. We have just got used to living with the unconscious, we know that all our habits have connections to hidden meanings. Nothing is psychologically innocent. Now we have to begin again, and recognize that nothing is ecologically innocent.[71]

This is a powerful example of the way in which the social context modifies and strengthens ecotheological texts.

Recognition of the specifics of the ecopathology of everyday life does not come easily. Myerson, quoting Ulrich Beck, argues that,

> in the face of ecological interpretation, people always pull back. The more revealing the insight, the more violent the recoil: "the resistance to insight into the threat grows with the size and proximity of the threat. The people most severely affected are often precisely the ones who deny the threat most vehemently, and they must deny it in order to keep on living."[72]

Is Myerson correct in his analysis? We think he is. There is no single proof we can point to. But the ecopathology of everyday life is becoming, at least for some, relentlessly hard to hide from—the carbon footprint of food production through factory farming, the second home, the dripping tap, the unrecycled plastics and paper, the contaminated soil that adds dramatically to the cost of building, the surge in ownership of private cars in China and India, the nonrecyclable Styrofoam cup, the air-conditioners that are powered by electricity from dirty coal—all may be reminders that planet earth is on the edge.

Hard as it may be to measure transformative impact, we argue that the ecotheological texts we are studying have been and are, in this new social context where nothing is ecologically innocent, part of what shapes a resistance to a social imaginary that ignores ecology. Even in an academic setting, it is hard to reduce these texts to just another SIM—take it or leave it. As texts for church and society, as critiques of the complicity of religion and the practice of consumption, as careful analyses of the way we live, as passionate pleas to hear the voices of women in the majority world, they seem to have matured from simply SIM to social movement, in a social context that is both resistant to seeing and cautiously aware of the ecopathology of everyday life.

3

IMAGINED FUTURES

The Christian texts of the mid-twentieth century expressed the convictions of Christian preachers and scholars—some of whom became prominent ecotheologians and are still identified with the field—that a response to the ecological crisis was urgent. Early ecotheological texts initiated the reformation of Christian teachings about creation and human responsibility for its integrity. One did not have to choose, they argued, between religious tradition and ecological sensibilities.

In terms of influencing the social imaginary, these texts enabled Christians to live their identities both as Christians and as environmentalists. Psychologists studying the relationship of ecological identities to sustained environmental action have discovered the strong role played by social affirmation in translating attitudes and values into action.[1] Ecological identity simply describes "how people see themselves in the context of nature, how people see animate and inanimate aspects of the natural world, how people relate to the natural world as a whole, and how people relate to each other in the context of the larger environmental issues."[2] A conclusion to one study of committed activist ecologists concluded that religions "stand to benefit by creating more welcoming spaces for their members with ecological identities."[3] This does not preclude the existence of multiple identities within one individual, but rather recognizes a newfound ecological identity.

In describing the way in which the social imaginary of the West grew from the interplay of ideas and practices, Taylor emphasizes the role of the imagination. The social imaginary is not primarily a set of "intellectual schemes" in which social reality is conceived in a "disengaged mode." Rather it is how people imagine their lives together, including what they consider normative. The normative values and images form the basis of expectations and practices that make up daily life in all its aspects.[4] Taylor also refers to these values, images, expectations and practices as the "available repertory."[5] Understood this way, the early ecological Christian texts sought to expand the repertory of Christian meanings, values, and practices to accommodate the emerging new ecological consciousness. They imagined a future in which to be Christian also meant to have an ecological identity. Furthermore, the early texts as well as those that followed showed consistent awareness of the place of religion within the social imaginary. They take modernity seriously.

EARLY ECOTHEOLOGY AND THE SOCIAL IMAGINARY

Christian ecological texts alone cannot create a brand new social imaginary in the West. Yet the interplay between these texts and ecological practices is likely to affect the social imaginary so that a nonecological social imaginary will have a diminishing moral legitimacy. Only those at some distant future can really judge if that is the case. However, it is clear that these texts contribute to a further development in the modern social imaginary.[6] Something new is being added to how people imagine their lives together. The Christian theologians who first addressed the ecological crisis were driven by urgency that all was not well in the social imaginary of the West. The ecological crisis was a result, they felt, of the way we have imagined and then lived our lives together. All were persuaded and sought to persuade everyone else that "the normal expectations" and "the common understanding" were inadequate and even for some, pathological. New norms and practices were necessary.

Some commentators and critics consider the ecologists of the mid-twentieth century to be retrogressive or even atavistic, calling for a premodern valuation of nature and a repudiation of modernity.[7] Textual evidence better supports the claim that the texts promote an advancement within the social imaginary—an enlargement of such key values as democracy, human agency, respect for difference, economic freedom, and accountable governance. What is imagined in these early texts in a large sense depends on these key values. The modern notion of humans as agents of history is a key assumption. The course of history must be changed, the texts argue, if the planet is to survive. It was precisely this assumption that was overlooked in the early negative critiques.[8]

H. Paul Santmire, in his comments on Wendell Berry (1968), called for an end to "the cult of the simple rustic life" within American religious views of nature. He was critical of the practice of "church camps" as the primary experience of nature for Christian urban dwellers. It was necessary, he argued, for the church to reconsider its theology of nature in the light of urban life and the sociotechnological realities in which Christians lived their lives.[9] These texts show an awareness of what was sacrificed in the creation of the social imaginary in its present form and call for a retrieval of some of what was left behind. At the same time the present must be dealt with in realistic fashion. Joseph Sittler expressed this sentiment as follows in commenting on Christianity's understanding of itself as historical: "The Christian believer is liable, therefore, to make an opposition, not just a distinction, between man-as-nature and man-as-history. This is the fateful separation which marks the post-Enlightenment community particularly."[10] His claim rests on substantial evidence of which he was quite likely aware. In 1934, Rudolph Bultmann clearly laid out what was a common stance of the leading theologians of the early twentieth century when he wrote:

> Our relationship to history is wholly different from our relationship to nature. Man, if he rightly understands himself, differentiates himself from nature. When he observes nature, he perceives there something objective which is not himself. When he turns his attention to history, however, he must admit himself to be a part of history; he is considering a living complex of events in which he is essentially involved. He cannot observe the complex objectively as he can observe natural phenomena; for in every word which he says about history he is saying at the same time something about himself.[11]

The theologians influenced by the events of the 1960s emphasized that the human is part of nature. They did not, however, wish to forego the achievements of those who had carefully and successfully built up the case for the human as part of history. Indeed, any action called for in response to the ecological crisis demanded a sense of human agency in taking responsibility for a new historical future. Hope rests on this assumption.

Joseph Sittler

Joseph Sittler expressed his first environmental concerns in a sermon in 1953 in which he included an excerpt from Fairfield Osborne's *Our Plundered Planet* that decried the modern human conception of nature as an inexhaustible resource.[12] Sittler also referred extensively to Rachel Carson's *The Sea Around Us*.[13] However, it is noteworthy that he claimed as the core source of his care for the earth his childhood in the rural Midwest as son of a Lutheran pastor. His education included an undergraduate degree with majors in biology and English. By his own admission, he had what is referred to above as an ecological identity. "It has never occurred to me," he wrote, "that my understanding of God could be threatened by galaxies or light-years. . . . I have never been able to entertain a God-idea which was not integrally related to the fact of chipmunks, squirrels, hippopotamuses, galaxies and light years!"[14]

Sittler wrote his first piece on theology and ecology in 1954.[15] It did not specifically address the ecological crisis, but did include discussion of the nature-history dichotomy in a critique of theology and its relationship to the culture of the day. His reference to W. H. Auden's *The Age of Anxiety* is telling in this regard. His critique stems from the sense of alienation he observes in modern humans, so that when he speaks of earth he includes the sociocultural context of human life. His pronouncements are profound and lay the groundwork for his later more intensive work on the ecological crisis itself. He writes, "This garden [Eden and thence Earth] he [man] is to tend as God's other creation—not to use as a godless warehouse or to rape as a tyrant."[16]

A key theme of Sittler's work is his focus on making theology relevant to the current world problems. His concern was primarily the reformation of theology. In other words, while there is no doubt that he wanted to address the ecological crisis and that he saw Christian theology as critical to that effort, he felt that Christians must have a deep meaningful basis on which to care for the earth. It was not enough to add new ethical injunctions; the core of theology itself needed renovation. He undertook to reinterpret some of the principal doctrines of Christianity, the relationship of nature and grace, and Christology.

As his biographers have noted, Sittler's teaching and life reflected the importance he gave to "little things," to the details of life. While he recognized the enormity of the ecological crisis, his teaching, preaching, writing, and living bore witness to the presence of grace in all of creation. Thus he sought to revalue nature and to turn his readers and listeners to a reverence based on the biblical notions of God's relationship to creation. In *The Care of the Earth*, first published in 1963, he traces from the Bible and the theology of Thomas Aquinas the distinction between enjoying nature and using nature to evoke a sense of the mystery and integrity of nature for "a world sacramentally received in joy is a world sanely used."[17] In *Nature and Grace* he more extensively reconstructs the theological meaning of nature that was lost to modern theology.

Sittler's treatment of Christology is really a long commentary on the cosmic Christ of Colossians 1:15–20. In his address to the World Council of Churches in 1962, another of the major influences on Sittler's thought becomes obvious—his involvement in ecumenism. He not only explicitly comments on the Christology of the Eastern Fathers of the Church, but also reflects on the fragmentation of Christianity, first as a division of East from West and then in further divisions during the Reformation. In the whole process of fragmentation in the West, "the realm of grace retreated" from the realm of nature. It was this movement within Christianity that permitted the "autonomous man" of the Enlightenment to claim the realm of nature.[18] Now, Sittler laments, we are "permitting to fall asunder what God joined together."[19]

The Pauline understanding of Christ and redemption, he claimed, was much larger than Christians presently understood it to be. Sittler insisted that what Paul was saying in Colossians was as follows: "A doctrine of redemption is meaningful only when it swings within the larger orbit of a doctrine of creation. For God's creation of earth cannot be redeemed in any intelligible sense of the word apart from a doctrine of the cosmos which is his home, his definite place, the theatre of his selfhood under God, in cooperation with his neighbour, and in caring relationship with nature, his sister."[20]

The primary theme of all of Sittler's work is the notion that it is a vital part of Christianity to care for the earth. Even if the ecological crisis did not demand it, the biblical message compelled the believer to respect and honor the creation. Following a comment on the obligation of everyone to respond to the ecological crisis, he says, "But the Christian is called, commanded, interiorly obligated to care for the earth *as part of being a Christian* [italics his]."[21] It was in this light that Sittler articulated the purpose of his efforts to re-interpret the traditional doctrines. He was concerned to relocate the doctrine of grace within the secularized world from which it had been extricated by modernity and its narrow interpretations of Christianity.[22]

Sittler's texts are mainly addresses, sermons, and essays rather than systematic treatises. They are as likely to appeal to poetry and art as to biblical and other traditional Christian sources. In secular art he found expressions of what he considered human perceptions of grace in nature. He defended his style of writing as itself a kind of rhetoric of grace. Like grace, it is relatively chaotic and unpredictable, but appropriate to a theology in the making. In Peter Bakken's description, Sittler's works are expressions of his own perceptions, not the well-reasoned and linear style of many academic theologians.[23] As was

Sittler's desire, his writings function as an invitation to further exploration of and motivation for responsible action in the world.

Sittler's works did not have a wide audience in his day. Because his first publications predated the larger response to the ecological crisis and even more perhaps because he is addressing committed Christians for whom the doctrines of nature and grace and the others he addressed would have resonance, he was virtually unknown to the wider public.[24] Yet his works form a compelling argument that to be a Christian is to have an ecological identity. Further, they support a praxis of respect and care for creation that is largely missing from and even antithetical to much of the modern ethos. Recent interest in Sittler's work reflects the coming-of-age of ecological theology.[25]

Thomas Berry

Thomas Berry was a contemporary of Joseph Sittler, although there is no evidence in either of their writings that they knew each other. They were, however, engaged with many of the same cultural questions and use many of the same references even though they were educated in different Christian settings; Sittler as Lutheran and Berry as Roman Catholic. While Berry did receive seminary education within the Passionist Order, most of his formal university education was in cultural history, not in theology. From the beginning of his academic life his interest was not primarily in addressing Christian theologians, but rather in speaking to the larger public. He did not exclude religious establishments, however, and was often invited to speak in primarily Christian settings. Indeed, the effect of his thought on Christian theology and practice is stronger and more obvious than that of Sittler.

Like Sittler, Berry claimed that his commitment to the natural world began from his childhood. In his 1999 essay, "Meadow Across a Creek," Berry reminisced about a moment when as an 11-year-old child he encountered a meadow full of wildflowers and grasses across a creek from his home on the outskirts of Greensboro, North Carolina. This became a defining moment and an experience that funded his life-long commitment to the earth. The meadow he refers to has long disappeared. The loss of the meadow illustrates for Berry the pervasive destructiveness of modern culture. He wrote:

> This early experience, it seems, has become normative for me through out the entire range of my thinking. Whatever preserves and enhances the meadow in the natural cycles of its

transformation is good; whatever opposes the meadow or negates it is not good. My life orientation is that simple. It is also that pervasive. It applies in economics and political orientation as well as in education and religion.[26]

The first public responses of Berry to the ecological crisis appear in the early 1970s. Themes that will later become more focused on the ecological crisis appear earlier. He is concerned with the large modern questions of human alienation and the relevance of religion to human progress, and addresses them in inter-religious contexts. During the 1960s, he joins the conversation about ecology as Rachel Carson, Fritz Schumacher, and others began to give it a serious public voice.[27]

Berry's insight into the cause of the ecological crisis and the way toward a solution is clear and sustained throughout his work. His seminal essay on the topic is "The New Story," first published in 1970. In that essay, Berry critiques the modern Western construction of the human-earth relationship, highlighting the role played by the Christian biblical tradition in the destruction of the natural environment. The ecological crisis is, he claims, primarily a cultural and religious crisis that will not be solved by technological fix-its. What is needed is a new way of understanding the place of the human in the cosmos—what he calls a new story. This comprehensive account of who we are, where we are going, and what we ought to do along the way is missing since the break-up of the medieval synthesis. This accounts for modern human alienation as well as the destructive attitudes of modern societies toward the natural world.

Berry was influenced by the works of Teilhard de Chardin, as well as by discoveries of contemporary scientists. He saw in the new account of the beginnings and evolution of the cosmos a promising basis for a new story that could support a functional cosmology, an account of everything that could ground an ecological praxis. He later collaborated with Brian Swimme, a mathematical physicist and student of his thought, in publication of *The Universe Story*.[28]

Berry was quite cognizant of both the gains and the losses of modernity. He brought to his texts a broad knowledge of the Western tradition and the development of historical consciousness. While not losing sight of the role of the human in the creation of history, Berry was passionate in his condemnation of the arrogance of Western culture in its disregard for and even hostility toward nature. He sided with those like Lynn White and Arnold Toynbee who saw the biblical tradition as complicit in the development of these attitudes and the consequent destructive practices of Western societies. While much of Western culture is now secular, and state and corporate decisions give little place to religious considerations, there are deep Euro-Christian roots in the institutions of Western society, he argued. Religion was part of the problem and now ought to be part of the solution. In order to contribute to the solution, traditional religions would need significant reformation and reinventing. Berry did not undertake this reformation himself except with broad stroke suggestions, such as taking seriously the discoveries of modern science and reclaiming some of the pagan earth sensitivities of the premodern era. Berry's texts, however, are a prophetic and impassioned plea for sanity and radical reenvisioning of the social imaginary. From the level of the human psyche that had become distorted away from its natural sources to the level of political, economic, social, and cultural institutions, Berry called for a reintegration of humans with the earth. "All human professions, institutions, and activities must be integral with the earth as primary self-nourishing, self-governing and self-fulfilling community."[29] "We now appreciate the more primitive stages of man's awakening consciousness. . . . These stages of human development have left an abiding impress in the depths of the human psyche, just as the various geological ages remain in the very structure of the earth."[30] In other words, Berry called for the reintegration of history and nature. In religious terms, this means an inclusion of the earth in our accounts of any spiritual journey. Commenting on the contribution of Teilhard de Chardin, he concluded, "The spiritual journey is no longer simply the interior journey of the individual soul or of the hero saviour personality, or the journey of the individual sacred communities or of the comprehensive human community, it is the journey of the earth itself through its various transformations."[31] Thus, while building on the great effort of Teilhard de Chardin to integrate understandings of the evolution of the universe within Christian theology, Berry goes further in presenting the ongoing evolution of the entire universe as itself a spiritual journey of which the human spiritual journey is one dimension.

Despite the sometimes-radical rhetoric of return to primal sensitivities, Berry does not suggest a return to the premodern or some postmodern repudiation of modernity. His ideas of advocacy on behalf of all of life and the granting of rights to all beings situate his work within what Taylor describes as the gradual extension of the modern social imaginary. In his 2004 Schumacher Lecture, "Every Being Has Rights," he argued forcefully that the rights accorded to humans in the U.S. Constitution while representing the "height of good aspects" of the modern world, were deadly for the rest of nature. All beings have rights by reason of their existence. These rights may be limited to reflect the nature of the being, but nevertheless the universe can function only when the right to integral being is recognized.[32] Like the gradual extension of rights since the Enlightenment to include more and more members of the human community, Berry calls for a revaluation of human rights to include a wider community of creatures.

It is difficult to assess comprehensively the extent of the influence of Berry's work. Among the more public and sustained evidence of his influence are the Elliott Allen Institute at the Toronto School of Theology, the Forum on Religion and Ecology, largely the work of his former students Mary Evelyn Tucker and John Grim, Genesis Farm, founded and run by Miriam MacGillis, a Benedictine nun and self-described disciple of Berry, the Centre for the Story of the Universe, founded by Brian Swimme, and individuals, such as Jane Blewett and Lou Niznik, who have dedicated large portions of their lives to promoting his ideas and fashioning their own lives according to them. The influence of Berry among green Roman Catholic nuns, for example, is documented by Sarah McFarland Taylor in *Green Sisters: A Spiritual Ecology*. She writes, "Overwhelmingly, the most visible theoretical influence on the green sisters movement has come from the work of the Passionist priest Thomas Berry. His work is indispensable to understanding the growth and development of an ecological ethos among green sisters in North America."[33] Many of the practices dealt with in chapter 7 owe their inspiration to Thomas Berry directly or indirectly through the formal and informal institutions his ideas have generated.

Rosemary Radford Ruether

Rosemary Radford Ruether's first ecologically conscious article, "New Women, New Earth: Women, Ecology and the Social Revolution," was published in 1975.[34] She had given a lecture of the same title in 1974.[35] Ruether speaks to the ecological crisis from her first area of concern, the liberation of women. She connects the domination of earth with sexism extending the thesis of William Leiss, who had connected the domination of earth with class domination.[36] Like other scholars, however, she is also provoked by Lynn White Jr.'s critique of the biblical sources of the ecological crisis. She attempts to correct what she considers his simplistic treatment of the biblical tradition. Ruether distances herself from others writing in the field by joining her concern for human liberation, especially the liberation of women, with that of ecology. She pleads: "Women must see that there can be no liberation for them and no solution to the ecological crisis within a society whose fundamental model of relationships continues to be one of domination. They must unite the demands of the women's movement with those of the ecological movement to envision a radical reshaping of the basic socioeconomic relations and the underlying values of society."[37] Ruether is recognized as the first Christian liberation theologian to relate the oppression of earth to that of women, especially in the majority world, as well as to other forms of oppression such as classism and racism.[38]

While her approach to the ecological crisis is different from that of the other early ecotheologians, Ruether does share with Berry the liberal and revolutionary spirit of post-Vatican II in the Catholic Church. She remains committed to the revision of theology and institutional religion to meet the needs of a contemporary and suffering world. She also has been deeply affected by the U.S. civil rights movement of the 1960s and 1970s. Ruether's own account of her family and formal educational background reveals an early awareness of the themes that highlight her work: an inclusive attitude that stretches the boundaries of her pre-Vatican II Roman Catholicism, strong female role models, and an attraction to the prophetic and liberation themes of the biblical tradition. The latter was responsible for her move from study of the Greek and Latin classics to a more concentrated study of early Christianity.[39] In 1964, Ruether worked with the Delta Ministry in Mississippi and experienced firsthand what she called "the black side" of the racial divide in the United States. From that experience, her 10 years of teaching at Howard University, a predominantly African-American college, her continued involvement in civil rights activism, and her exposure to Latin American Liberation theology, Ruether's life work has deep roots in the realities of domination and oppression. These are the recurrent themes through which she understands and explains the interconnections of race, class, and sex in her ecofeminist theology. Hers has always been a theology of liberation.

In the preface to her edited volume, *Gender, Ethnicity, and Religion: Views from the Other Side*, Ruether's own words are explicit in indicating the multiple facets of her theology represented by the voices she brings together: "Finally, the volume also offers new theological perspectives, arising from these many contexts—Afro-Caribbean, Cuban, womanism in the United States—as well as insights into sacramental theology and spirituality arising from the many challenges of ecological crisis, the lives of people with disabilities, and the civil rights struggle."[40] In all of these areas, Ruether's thought is consistent in highlighting the ethos of Western domination resulting in large measure from a dualism that favored history over nature, the human over the natural world, the soul over the body, man over woman, reason over religion, and the like. Her vision of a future is this-world centered, but not anti- or premodern. She calls on the liberative forces of modernity, but with an eye to what has been neglected or abused. Her theological anthropology is a new understanding of human rationality in which the human is perceived as a creature of earth whose approach to the other creatures ought to be one of care and respect. Ruether assumes that reason is a critical component of humanness. However, she critiques its narrow Enlightenment understanding as one term of a dualism in which it is separate from and dominant over emotionality, the body, and all other components of creaturely existence.

Ruether's ecofeminism emerged within the formative debates among ecofeminists as they attempted to theorize the base of their concerns. Were women inherently more ecological, closer to nature, than men? Ruether's early and consistent stance was that whatever differences existed between the male and female roles in nature were explainable by the socialization of women to tasks connected to home and local land. Like other ecofeminists and many other ecologists, Ruether seeks to re-enchant life and the natural world away from the instrumentalist and rationalistic thinking of modernity. However, she is not advocating a return to a premodern state.

Ruether's life project can be understood as an invitation to what Chaia Heller describes as a re-enchanted rationality. Using the language of social desire, Heller proposes a kind of rationality that includes a passion for nature, for connectedness to others, and for social justice. "When disenchanted by a rationalized and 'McDonaldsian' world, we confuse rationalization with rationality and look immediately to intuition and the spirit for both solace and a solution."[41] In her conclusion to *Gaia and God*, Ruether calls for communities that incorporate both new liturgical expressions of connectedness to earth and to one another as well as a critically aware resistance to the deadening and unjust structures of modern life. These are communities of engaged and passionate rationality; in her words, communities of "celebration and resistance."[42]

While drawing on the best of Western Christian and modern bases in human agency, solidarity, empowerment, and human rights, Ruether's vision is also radical in the degree to which she pushes the boundaries of Christianity. The radical nature of her ecofeminist vision is often masked by her ability to remain a part of the Roman Catholic communion.[43] She consistently critiques not only the anti-feminist beliefs and practices of her own institutional religion, but also of the deep doctrinal understandings that in her view have grounded oppressions of all sorts, including women and the earth, especially as these affect the poor.[44] Her departure from the classic Christian doctrines of eschatology reveals the extent of this world- and earth-centeredness that she envisions. With regard to death, she postulated, "our existence ceases as individuated ego/organism and dissolves back into the cosmic matrix of matter/energy . . . into the great organism of the universe itself."[45] In body and spirit we are of this earth, both in life and in death. While Ruether's contribution has been primarily in her efforts to change thinking and reformulate Christian traditions according to the more liberating and healing aspects of those traditions, she, like the other ecotheologians whose texts we are presenting, is not content with the transformation of consciousness. She argues passionately that "the transformation of consciousness is the servant of a struggle to transform this entire social system in its human and ecological relationships."[46] Her texts encourage and provoke such a transformation.

John B. Cobb Jr.

John B. Cobb Jr. describes a conversion experience in 1969 that set him on the course that makes him one of the earliest and most prolific ecotheologians. Cobb had already related theology to the issues of the twentieth century by adopting the new and developing process theology as his field of study. By his own account it was his teenage son, Cliff, who prodded him to think more deeply about the ecological crisis. Finally in the summer of 1969, the elder Cobb reports, "The issue of human survival seemed so overwhelming in its importance that I felt I must reorient my priorities at once."[47] His first response was to organize a conference on the issue. Out of that conference, Cobb published *Is it Too Late? A Theology of Ecology*.[48] Reflecting on his experience later, Cobb wrote, "It seems then as it seems now, that we must have images of a hopeful future."[49] From that realization, Cobb began and continued to work with activists trying to bring about change, but primarily he created texts that he was convinced were necessary for the transformation of attitudes, lifestyles, and institutions. He remains convinced that a rethinking and a reformulation of Christianity is a key element in that transformation.

Cobb engaged the philosophy of Alfred North Whitehead and the theology of Charles Harshorne toward a new and systematic formulation of the bases of Christian theology. Christianity, Cobb argued, must engage all areas of human research and development, physical and social sciences, politics, liberation movements, and other issues of the times. To encourage the ongoing conversations among representatives of all these concerns, Cobb, together with David Ray Griffin, established the Center for Process Studies in 1973.[50] Cobb's work reflects his commitment to the future of Christianity as a holistic system, one that unifies ongoing intellectual endeavors so as to present a firm basis for Christian involvement in all that matters in the world. The other side of his concern is to provide the scientists with a language and a space within which to reflect on the ethical and social consequences of their day-to-day work. The titles of his published works reflect his various concerns and overriding commitment to provide a blueprint where Christianity can contribute to a more just and sustainable world. In *Reclaiming the Church*[51] he wrote of the necessity of rescuing Christian churches from trivial and meaningless formulas and practices to a revitalizing involvement with the major issues of the contemporary world. *For the Common Good*[52] (written with Herman Daly), *Sustainability*[53] and *The Liberation of Life*[54] explicitly reflect his primary concern with the ecological crisis. His commitment to inter-religious dialogue resulted in a volume edited with Christopher Ives entitled *The Emptying God: A Buddhist-Jewish-Christian Conversation*[55] (1990), among others. A collaborative spirit and an outreach to the contemporary world for engagement and conversation characterize all of these texts. This is Cobb's vision of a sustainable future world.

Cobb's commitment and passion for meaningful change toward a more ecologically sustainable world is demonstrated by his tenacity in confronting the difficult questions that plague and frustrate those long involved in this issue. Why, Cobb wonders, is it taking so long despite a developing consensus among Christians regarding the urgency for change, for real and substantial change, to happen?[56] He addresses the question by examining the intricate relationships between the technological and economic practices of North American culture and the often lofty aims of Christians to resolve social problems such as poverty. As Christians we continue to buy into the claims of our present economic system and its emphasis on growth and progress as the solution to poverty through development. Addressing the Christian churches at the close of his reflections, Cobb states, "The constructive ecclesial task is both long-term and immediate. The pressing immediate task is to address one-by-one the downside effects in city neighbourhoods of the global economy's impacts and to try to find, in practice and policy, those incremental changes that slowly build toward sustainable communities.... The long term task, succinctly put, is the conversion of Christianity to Earth."[57] By conversion to earth he means the overcoming of our modern estrangement from the natural world. He means recovering the power to recognize and experience the mystery and agency that extends beyond the human community and that derives from the presence and power of a God incarnate in the entire universe.

As is the case with other early and long-term ecotheologians, Cobb's life work holds up a hope that reaches from the lofty aspirations of religious consciousness to the practical embodiments of these dreams for a better world. His texts challenge ever more strategically the interweaving of hope and practice. Like other ecotheologians, he calls for a recovery of sensibilities to the natural world that were cast aside or simply forgotten in the evolution of the social imaginary in the West. Under the horizon of the ecological crises apparent in the mid- to late twentieth century, he imagines and models a future that is a further extension of the modern social imaginary. Clearly there is a future, and hope rests on practicing it now.

4

THEOLOGY AND THE ECOLOGICAL CRISIS

There is a crack in everything.
—Leonard Cohen

Christianity played a significant role in the creation of the Western social imaginary. Whether one is referring to the impulses toward the ideals of the enlightenment or the negative impact of Western civilization on the ecology of the planet, the case has been made for strong biblical and Christian bases for Western development. But Christian believers also see history through a theological lens, and this chapter reflects on the confessional stances of Christian ecotheological texts. In the language of Christian theology, history is the realm of redemption. The dynamics of the formation of a certain social imaginary are subject to judgments about what is clearly liberative and redemptive and what is not. How *ought* the social imaginary develop? What is not yet liberated is more than just a stage in a continuously evolving social imaginary. Theologically it represents a state of decline, a culture in need of redemption. Canadian theologian Bernard Lonergan used the terms progress and decline to refer to the dynamics of human history. His clear articulation of a theological understanding of history is helpful in appreciating the significance of ecotheology, which attempts to engage the ecological crisis from a Christian theological perspective.

Some theologians were deeply troubled by the accusation that the Bible and the Christian tradition were in large part responsible for the degradation of the earth. They delved into the tradition to reexamine, retrieve, reinterpret, and reform the texts and practices responsible for such devastation. They understood the drift into the ecological crisis, and used theological tools to articulate what went wrong and how best to right it.

As claimed in the previous chapter, these theologians were not really creating a new social imaginary. Many of the gains of the post-Enlightenment imaginary were called upon and reexerted in response to the ecological crisis. Primary among those was the sense of human agency in history and the requirement intentionally to push back all forms of colonization, imperialism, poverty, and oppression, all of which would come to be seen as intricately tied to the ecological crisis. Christian theologians who took the ecological crisis seriously now viewed the evolution of Western society in the light of the ecological crisis.

To what extent was Christianity responsible for a greatly degraded creation? How could and ought Christian people lament the mistakes of the past and move forward, and creatively contribute to the healing and sustaining of all creation? There are a variety of responses. Those we consider below do not include responses that hold either (1) that there is no credible evidence of ecological crisis or (2) that there is such evidence, but this is not of Christian concern because the world will pass away; that we are heading for an apocalyptic transition to a new and better other world.

The theological stances we consider are those that choose to enter into serious dialogue with the various dimensions of the ecological crisis and to respond from the resources of theology. The theologians concerned all recognize some degree of complicity. Certainly all are convinced that Christians must not only play an active role in addressing the crisis, but that theology must be articulated in a new way in light of this responsibility. In other words, like Charles Taylor, ecotheologians are convinced that the creation of a reformed social imaginary requires interplay of new and innovative ideas and determined action, of new meanings and practices. Their project is none other than to change the ways in which human communities inhabit the earth. Their focus, however, is on the Christian community's role in doing so.

THEOLOGY IN LIGHT OF THE ECOLOGICAL CRISIS

Most ecotheologians begin their work with a litany of ecological woes or at least with a reference to someone else's litany. This is particularly true of the first theologians to respond, some of whose texts we have discussed earlier. Some also lament the growing irrelevance of theology to contemporary concerns in general. All view history from a theological perspective. What does this Christian theological perspective add to Taylor's understanding of the social imaginary? Obviously, it adds faith in a divine dimension to history, a presence of an active and caring God who has already redeemed the world. Indeed, a large part of the work of the ecotheologians is a strenuous reclaiming of the physical world (earth or entire cosmos) as a sphere of God's action and caring. For that reason, it is important for ecotheology to have a theological account of what has gone wrong. Simply put, the conclusion is that ecological degradation is not divine intent, but that it results from the gradual accumulation and systematization of human self-interest, lack of attention, and a long list of other human failings that created a drift in our present social imaginary.

Bernard Lonergan's account of history describes this drift theologically as a longer cycle of decline.[1] Most ecotheologians do not explicitly make use of his account, but it is helpful for its heuristic value. Taylor's notion of the social imaginary is akin to Lonergan's notion of common sense.[2] Like a social imaginary, common sense exists everywhere, but it is shaped quite differently in particular places and contexts. Both social imaginary and common sense refer to the ordinary life of citizens, the day-to-day practices that integrate versions of high-blown ideas, useful practices, ritualized ways of communication and various other resources that imagine a viable and fulfilling society. Both evolve and change over time for better and for worse according to what works best in the short term. Sometimes there is revolutionary change. As power changes hands, the old imagination is no longer useful, and a revised social imaginary, a new common sense, takes hold.

In this kind of account of the development of the West, ecological awareness was not a key component. Commonsense life did not perceive ecology's relevance and by many accounts moved further and further away from attentiveness to the plight of the natural world. Nature was a resource and not much else. In Taylor's terms, this would seem to constitute part of the drift of the social imaginary; in Lonergan's terms, it is indicative of the longer cycle of decline.

Lonergan's account and Taylor's are not exactly the same, and the differences are a precise indicator of what Christian theology offers to the understanding of history. Taylor's account is a description of how things are and came to be. Lonergan adds a theological analysis of how things ought to be, given the reality of redemption and the action of a loving God mediated by the converted intentionality of humans. For Lonergan, as for the ecotheologians we study, this is not an automatic or easy task. It calls for the consistent difficult attention to what Sittler referred to as God's "theatre of grace," in a world always on the move as horizons change. It is the assumed basis for the ecotheological texts we are examining and the underlying justification for declaring them practices of hope.

The question of the relationship of theology to culture is a problematic one, nonetheless.[3] How does theology relate to culture and what really do theologians mean when they make the claim that theology is a cultural construct? Does theology belong primarily to culture understood as high culture, the acumen of intellectual or spiritual achievements of history? Or does it belong primarily to culture as the everyday life of Christians in a particular cultural setting? Which does it seek to elucidate and for what purpose? These are ongoing questions. However, for the purposes of the discussion of normativity as we employ it in this discussion, the assumption is that theology exists for the wider Christian community to bring meaning to its practices and beliefs. As a practice itself, academic theology may make use of the resources of high culture as well to elucidate existing practices and beliefs, messy as they are, and to challenge Christians to new social practices, for example.[4] In Kathryn Tanner's words:

> We mean that the problems that academic theology tackles are prompted by social practice. We mean that academic theology is about what day-to-day Christian social practices are also often about; for instance, the propriety of new proposals for Christian belief and action, when previously established Christian practices are challenged by new situations.[5]

In the end, however, no theologian would claim that his or her theological proposal is merely a suggestion about change. Certainly no ecotheologian discussed in this volume would. In Tanner's own proposal (as indeed in Lonergan's) the performance of the theologian herself and the application of judgment plays a large part in determining how the implicit ought (about what is to be done) is constructed and offered to the wider community. But there is always an ought no matter how tentative or contested and even if it is addressed to a culturally specific community.[6] In summarizing recent investigations and insights on the relationship of theology to culture, Tanner holds for the role of academic theology as a service to everyday life, much as Lonergan argues for the critical role of theory, in general, to common sense.

Lonergan observes that there is an ambiguity within the sphere of common sense. While the strength of common sense is its attention to the here and now, to the practical concerns of life, this can also become its limitation. This happens when the society ignores the appropriate roles of higher explanatory viewpoints characteristic of theory. Thus there is no comprehensive view on which to base human decisions and activities. Explanations characteristic of theory are the result of reflection on historical realities and are a requirement for wise judgments about what ought not be repeated or what ought to be enforced. Longer views can preclude what appears for the moment to be practical but would be detrimental in the long run. Common sense, in denying the value of theoretic viewpoints also consciously refuses insights deemed impractical or contrary to immediate interests.

As Lonergan was aware, practical realism is necessary and it acts as a corrective to theories that ignore the ground of life experience. Over the course of time, however, a constant and systematic omission or refusal of more comprehensive explanatory views results in a process of decline. To return to our example of academic and everyday theology, without the resources of academic theology adeptly and compassionately interwoven with everyday Christian belief and practice, Christian life is likely to miss out on the "critical and evaluative questions."[7] How consistent is a Christian belief in love of neighbor with the practice of discriminating against one's neighbors of color? How does Christian acceptance of the goodness of a universe divinely created allow for indifference to the abuse of nonhuman creatures? In a word, theology has a role to play in the resistance to decline. Ecotheologians construct their theologies in resistance to a decline which finds expression as ecological crises.

ELEMENTS OF DECLINE AND CHRISTIAN RESPONSE

Ecotheological texts were constructed in the light of ecological responsibility. Berry's and Merchant's accounts are different, but complementary, accounts of the complicity of Christianity in ecological decline.[8] The interaction of ecotheology with other theologies, such as liberation theology and feminist theology, expanded the accounts of decline to include the role of economic expansion and globalization, racism, and the persistence of poverty and oppression, especially in the majority world. In doing so, these theologians engaged the resources of the social and physical sciences, economics, history, and other relevant disciplines. The task was to situate theology firmly within this world and to call humankind back to its earthly abode. This would require the creation of new meanings and values in order to correct and move forward the social imaginary in light of ecological responsibility.

Ecotheology Meets Social Justice

The ecological movement took most of its impetus from the activism of the 1960s and 1970s. In the social space created by those movements, as well as in the academic spaces of universities and the updating reforms of Roman Catholic and other Christian churches, ecotheologians found their roots. This interplay of imagination, action, and reflection continues to be a hallmark of ecotheology as it widens its scope.

The charge to ecotheology to consider social justice paralleled a similar movement in the wider society. Giovanna Di Chiro relates the story of a south central Los Angeles woman, Robin Cannon, who helped lead a resistance by the mostly African American citizens to the location of a 1,600-ton-per-day solid waste incinerator in the center of their neighborhood. By forming a coalition of community organizations and local business and community leaders, she and a few other women successfully led the effort to block the construction of the waste incinerator.[9] Di Chiro reports surprise that environmental groups such as the Sierra Club and Environmental Defense Fund did not support this effort in the beginning, at least, because they identified the issue as community health and not environmental.[10]

This distinction and virtual separation of concerns about the natural world and concerns for human communities was a common (if not universal) characteristic in the early evolution of the movement. The conversation between Gregory Baum and Thomas Berry recorded in *Thomas Berry and the New Cosmology*[11] reflects this critique. On the other hand, the social justice or liberation theology of the time was silent on ecological issues. When the renowned liberation theologian, Leonardo Boff, wrote *Ecology and Liberation: A New Paradigm* in 1993 (English translation 1995), the book was welcomed as offering a much-needed perspective to ecotheology.[12] Moreover, it was the addition of a much-needed perspective to liberation theology.

Despite the perception that the environmental movement was silent on social issues, it is worth recalling that the roots in the mid-twentieth century were tied to anti-nuclear campaigns. Perhaps even more significant was the effort of Cesar Chavez and the United Farm Workers (UFW) to gain more protection and rights for farm workers from the grape growers of California. Based on insights of Rachel Carson's *Silent Spring*, the workers and their supporters lobbied against the use of DDT and other toxins on union ranches, among their other concerns. Although it took at least 10 years before the Environmental Protection Agency prohibited the use of DDT (longer for other toxins) on farms employing manual laborers, the example of this mostly Latino effort illustrates the way in which ecological, health, and social issues were seen as mutually implicated by many of those people most affected by both.

According to some environmental historians, the action that galvanized environmental activism and put environmental racism in the forefront of the environmental movement was the demonstration in Warren Country, North Carolina, in 1982. The residents, predominantly African American women and children, used their bodies to block trucks filled with Polychlorinated biphenyl (PCB)-laced garbage destined for a landfill near their communities. Several other demonstrations followed in other parts of the United States.[13] The effectiveness of the efforts, not only in some cases achieving their local goals but also in attracting public attention, was enhanced by a series of studies to determine the extent that minority and low-income communities suffered disproportionately from environmental toxins as a result of landfill and dirty industries. The first of these was conducted by a Christian denomination.

In 1987, the United Church of Christ in the United States conducted a study to investigate the relationships among income, race, and exposure to environmental toxins. The study concluded that while level of income was definitely a factor, the minority status of local inhabitants was a much larger predictor of the location of commercial hazardous waste facilities. In the words of the report, "The proportion of minority members in communities with commercial hazardous waste facilities is double that of communities without such facilities. Where two or more such facilities are found, the proportion of minority members is nearly triple that in otherwise comparable communities. In fact, the best predictor of where to find hazardous waste is to classify communities by race, not income or real estate values."[14]

The director of the investigation, Benjamin Chavis, coined the phrase "environmental racism" to describe this reality.[15] Other reports corroborated this finding not only for the United States, but also internationally. In 1988, Nigerians and other African people were becoming vocal about the dumping of toxic chemical waste, including material containing high levels of PCBs, in their countries. They viewed this action as just another installment of the "historical traumas" in line with colonialism and the slave trade inflicted on Africa by Western nations. An editorial in *West Africa* referred to the incidents of waste dumping as "toxic terrorism."[16]

Furthermore, multinational corporations promised development and economic benefits to poor countries without the power to insist on worker rights and environmental protection for their land and peoples.

While globalization may have brought benefits to some, the global drive for wealth was seen to accelerate ecological problems around the world. Toxic dumping was not some renegade criminal action; it was part of a global system of agreements among nations. As the case of Nigeria demonstrates, the actions were supported by the World Bank and problems stemming from oil exploration were often part of collaboration between executives of large corporations and corrupt and oppressive governments.[17]

Besides presenting a dimension of horrific injustice in its own right, environmental racism emerged as a category in environmental and theological contexts (UCC) also illustrates preeminently how ecological devastation, while evident universally on the planet, has different effects in different contexts. In addition, it revealed the complex interrelationships of systems of oppression of people, especially poor and minorities, and devastation of the natural world. Whether justified or not, the ecological movement was being called to task for a monolithic critique of the forces of natural destruction and proposed solutions that ignored the complexity of power and unjust human interactions around the world. As Giovanna Di Chiro reported in her contrast of environmental justice activists and mainstream environmentalists, "They [activists] therefore contend that the mainstream environmentalists' invention of a universal division between human and nature is deceptive, theoretically incoherent and strategically ineffective in its political aim to promote widespread environmental awareness."[18] Activists for environmental justice, who were predominantly people of color, declared themselves the "new environmentalists."[19]

Ecotheology and Corporate Expansion

While some Christian ecotheologians may be accused of focusing on the recovery of respect and reverence for the natural world to the neglect of social justice issues, most did not take a radical ecological stance that ignored human concerns. As early as 1975, Rosemary Radford Ruether was making connections between human domination of other humans and human domination of the earth. Many ecotheologians were no doubt cognizant of this expansion of corporate power and its potential dire implications for the world. However, the question of the alleged extreme anthropocentrism of Christianity required attention, and liberation theologies to that point did not seem cognizant of this distortion in the tradition and its effect on the natural world.

The Christian tradition had an admirable history of social teaching and action, but had exhibited indifference and sometimes even contempt for the natural world. Hence, a lot of effort was expended in reconstructing that relationship theologically. The edited volume, *Earth Might Be Fair*, published in 1972, is an example of such an effort.[20] The articles in this volume all focus on the human-earth relationship, bringing together the latest scientific understandings with ethical and religious teaching. Writing his critique and defense of the Christian tradition in 1994, Richard A. Young mused that "Perhaps the slowness of the Christian community to become environmentally active is due to the slowness of theologians to formulate a theology of nature."[21] This view was shared by many.

Some saw the formulation of a theology of nature to be a difficult and all encompassing task given the heritage of disregard and even contempt for the natural world inherent in much of the Christian tradition. Catherine Keller calls such attitudes a particularly "thorny obstacle" in the way of a theology of nature. Protestant Christianity, she claimed, has "denaturalized" the world, seeing material reality as seductive, likely to inspire idolatry, and hence creating an abstract, individualistic and other-world focused spirituality. Not only does Christian theology militate against ecological responsibility, however, but in its form as Christendom was the carrier of the colonization of the planet, both human and nonhuman.[22] Furthermore, in later work, Keller argued that the very understanding of the Genesis accounts of creation has been colonized by Christian dogmatic interpretation. The doctrine of *creatio ex nihilo*, traditionally held by most Christians, she claims, is not supported by the biblical texts; the static notion of origin inherent in the Christian doctrine replaces that of beginnings. Origin denotes that which is absolute whereas beginning is always "relative, contested and historical."[23] In the light of findings on environmental racism and the growing awareness of the particular and diverse ways in which the ecological crisis was affecting different contexts, the critique of global systems of oppression become more sophisticated and more and more a hallmark of ecotheology. Even so, the concern with anthropocentrism continues and is brought very much to the fore in *The Earth Bible*, which is a sustained and compelling effort to reread the biblical texts from an earth perspective.[24]

Sallie McFague's *Life Abundant* is one example of an ecotheology that sees a long cycle of decline associated with growth of corporate economic control of the world. North American Christians in particular, she claimed, are complicit in the dynamics of oppression and ecological degradation associated with an economic system gone awry. Beginning with the same movements toward individual freedom, classical economics, and industrialization from which Taylor outlines the construction of the social imaginary in the West, McFague traces the darker side. She admits that there were advances in freedom and well-being for all, but focuses on those left behind as well as on the consequences for the natural world. According to McFague, a worldview emerged in the West that she describes as "a worldview of mechanistic progress."[25] This worldview arose in the seventeenth century and had its source in the industrial and scientific revolutions. She also links it to the Protestant reformation with its emphasis on individuality. All of this brought vast improvements in quality of life for many people; it is, however, becoming "ragged at the edges as rumbles of climate change, AIDS, poverty and population surges, deforestation, desertification, and overfishing, and so on increase, but for most North Americans denial is still possible."[26] A constructed worldview, McFague explains, is generally taken as "natural" and universal; it permits those who hold it to go about their daily lives as if things are as they are and not significantly changeable. It works for those who hold it. Questions such as "Who is benefiting [from this worldview]? How long can this upward line [of progress] continue? What are the consequences for the planet?" do not ordinarily arise.[27] The worldview acts as a mask for its consequences; namely, the suffering of many people in different contexts on the planet, as well as the degradation of the planet itself. Both of these consequences, for people and for the planet, are linked to a pervasive worldview that, when examined, reveals the lines of its own construction as well as the sources of its power.

The worldview of mechanistic progress combines the notion that the earth is a machine with the notion that it is possible to produce more and more "goods" for more and more people *ad infinitum*. Thinking of the earth as a machine enhanced greatly the ability of scientists to manipulate the earth, to take control of its parts, and to develop technologies from its resources without remorse or sense of limit. The planet had no sensibility, no fragility, and no inherent sacredness. Human history on the planet was pictured as a continuous and endless line upward, where "up" has come to mean not only an extension of human rights and healthy quality of life, but also increasing availability of consumer goods.

McFague calls attention to the development of a "neoclassical" economics that has become integral to this worldview. This recent and contemporary expression of economics is characterized by a claim to neutrality, an anthropology that defines human action as motivated almost exclusively by self-interest, and the assumption that an "invisible hand" will guide the competing markets in the best interest of the most people. In other words, "The key concept in neo-classical economics is the argument that freely acting, acquisitive individuals will eventually, though not intentionally, work out the best solution for production and consumption in a society."[28] Neither distribution of profits from the earth's resources nor the ability of the planet to sustain such an economic system is part of neoclassical economics. Such concerns require intentionality, a constant reevaluation and reconstruction of the economic system, an acknowledgment that economics as practiced is value-laden, and that other worldviews exist. Implicit in this view of economics is the supreme importance of the individual. In McFague's view, Adam Smith capitalized on a version of Christian Protestant anthropology that gave freedom from church and state to the individual, but also emphasized human sinfulness. Leave the market to individuals and rely on their sinful inclination to greed and acquisitiveness!

McFague is not the only ecotheologian to construct her version of what we are calling (following Lonergan) "the longer cycle of decline." Most versions of how we came to be where we are socially and ecologically are similar.[29] All see the process as a complex interaction of many factors, most unintentionally contributing to the present crisis, but all subject to theological reflection, critique, responsibility, and change.

GLOBALIZATION, POVERTY, AND THE ENVIRONMENT

The ecotheology emerging in the late twentieth century and early twenty-first century has been characterized by broader social critiques than the earlier efforts. Ruether's summary under the chapter title "Corporate Globalization and the Deepening of Earth's Impoverishment" integrates most of the elements of the critique. Like other ecotheologians, she is in conversation with the works of environmental scientists, feminists, liberation theologians, social scientists, economists, and representative voices from non-Christian traditions, particularly with voices from the majority world.[30] Because of her long-term efforts to provoke and encourage Christian reflection on many of the seemingly intransigent ecological and social problems of the modern world, Ruether can outline the interconnections of the historical inheritance contributing to these problems, as well as the newer drifts and elements of decline evident in a postcolonial world.

"For me," Ruether writes, "what is being discussed today as 'globalization' is simply the latest stage of Western colonialist imperialism."³¹ She divides the colonial period into three stages: (1) from the late fifteenth century to the independence of the American colonies; (2) from the mid-nineteenth century until the 1950s, a period that included the dividing up of Africa, much of Asia and the Middle East by European countries, the formation of the British Empire, and the intensification of colonial efforts by war torn European nations after World War II; and (3) a neocolonial period beginning in the 1950s when many of the colonized nations were granted "flag independence," but in fact most of the power was handed over to white European descendants within those nations. The emergence of the United States as the greatest military power after World War II, the existence of communism and its perceived threat by the United States and its allies, and the continued impoverishment of most of the nations of the world were key elements in a context in which capitalism as an ideology was associated with freedom and held up as the answer to the world's problems. Ruether concludes: "By demonizing communism as atheistic totalitarianism, and pretending to be the champion of 'democracy,' the West masked the fact that what this crusade was mostly all about was the maintenance of neo-colonial Western-controlled capitalism and the prevention of genuine locally controlled political economic democracy."³²

Globalization is really the third phase of colonialism. Globalization is characterized also by the intensification of efforts to "develop" the poorer nations. As Europe recovered from World War II, institutions such as the World Bank, International Monetary Fund (established 1944–1947), and the World Trade Organization (since 1955), which were concerned primarily to rebuild Europe, focused on "loans" for development of the poorer nations. These loans were funded by contributions from the member nations who also control decision making, including who receives loans, and under what conditions.

Among other dire consequences of this powerful system is structural adjustment. Poor nations defaulted on the loans; projects for which borrowing was done were never completed. For various reasons, including unstable new governments, dictatorships, and general poverty, borrowing nations could not repay the loans. A debt crisis resulted. Structural adjustment was instituted. Borrowing nations were required to adjust their internal laws and institutions to enable timely repayment of the loans together with interest. (In most cases, interest is about all that the borrowing nation can hope to repay in the foreseeable future.) This adjustment generally includes reduction of social programs, increased privatization, and a major reduction in trade barriers to enable foreign investment.

Ruether lays out some of the particular impacts of structural adjustment programs; generally, they incurred greater poverty and misery for those already suffering economic and social oppression.[33] Despite this increase in misery for many, local elites prospered, as did multinational corporations. With the minimization of trade barriers and subsidies from their national governments, the whole world was an accessible resource, both as raw materials and as cheap labor, as never before. The growth and availability of sophisticated technologies contributed also to economic globalization.

The severe gap between rich and poor as well as other factors, such as the stringent international expectations placed on national governments with regard to basic services to their populations, contributed in large fashion to ethnic conflicts around the world. Even more significant, Ruether claims, is the attempt (often violent) by rich nations such as the United States to ensure the survival of regimes around the world that are favorable to the agendas of rich nations. Often under the guise of promoting greater democracy, what is really promoted is a climate favorable to economic growth in the already rich world.

Finally, Ruether outlines the implications of globalization as she describes it for the environmental crisis and for women in particular. She cites as examples: climate change, air pollution, industrial agriculture, privatization of water, and population growth.[34] The following excerpts from her account of industrial agribusiness indicate the intricate relationship she identifies among the social, political, economic, and ecological contexts of the present world crisis: "The patenting of seeds and its defense as 'intellectual property' of seed companies become a major expression of 'biopiracy' of communities' rights to their traditional foods. Global corporate agribusiness is causing a major crisis in the future of the entire human food supply, or what Vandana Shiva and others call the 'hijacking of the global food supply.'"[35]

Similarly, speaking of the role of the World Trade Organization in the unjust distribution of water because of increasing privatization, she reports, "In many cities in Latin America water is delivered at reasonable cost to the taps of wealthy households, while the urban poor buy it by the can from private water carriers who may charge as much as 100 times the rate of the city services."[36] In the meantime, "The water corporations [such as Vivendi and Suez] seek political power over local and state governments by sitting on international business councils and inserting a bias in favour of privatization of water as a commodity into trade rules upheld by regional trade treaties, such as NAFTA, and bilateral investment treaties between countries and the WTO."[37]

Similarly for the other ecological issues, Ruether draws on existing reports and scholarship to paint a picture of a complex and dynamic interaction of human motivations and systems that have contributed (and continue to contribute) to an unjust and unsustainable world. In her treatment of the gendered nature of unjust and unsustainable situations, she also points to the complicity of religious ideologies (such as right wing Christian influence on U.S. policies on population) in the suffering she describes.

The lines of Ruether's critique of globalization are not unique to her or to the others who construct similar pictures of an unjust and unsustainable world.[38] It is the dark side, the drift, of the Western social imaginary as described by Taylor. More particularly, it is the longer cycle of decline operative in aspects of recent human history; hence, to follow Lonergan, it is the arena of redemption. Ruether's (and others') accounts give explicit content and context to the contours of the longer cycle of decline with the implicit judgment that this is not what ought to be; neither is it the last word. But any notion of redemptive word or action by any theology will not be authentic or have any chance of being effective unless the particular realities are known, understood, and addressed.

Furthermore, and what is specific to the ecotheology texts constructed by Ruether and others is that the earth counts. Redemption applies to the earth community, the fate of the water and soil, as it does to the fate of those who depend on the earth for their survival. With all the same awareness of the socioeconomic and political ills outlined above, Norman Habel observes in his introduction to the *Earth Bible Series*, "The effect of this thinking [seeing creation themes as mere insertions in a story of redemption in the Bible] was to support, perhaps inadvertently, the devaluation of nature as God's domain in favour of human history as the arena of God's mighty acts of salvation."[39] Ecotheology texts more and more have spoken of this arena as one in which earth and history mutually constructed each other and one of which theology must take serious account.

BREAKING THE CYCLE OF DECLINE: TRANSFORMATION

In Lonergan's theological terms, breaking the cycle of decline involves a process of conversion of the whole person understood as person in community. This is a transformation of consciousness—the intellect, the will, the psyche—which ultimately relies on the love of God that floods one's heart and translates into authentic action for the resistance of evil and the promotion of all that is good. This is not an easy transformation; it involves the constant struggle to pay attention to the world, observe it intelligently, listen intently, and to be responsible and loving as we judge and act in relation to relevant horizons.

That, in effect, is the struggle of the ecotheologians to lay bare the intricacies of the ecological crisis and make judgments about what might be effective action for positive change. They do so as theologians. So the transformation of the self, of community, of society and culture is understood in theological terms. A key question for many as they sought to expand their traditions to incorporate the earth in serious fashion was: What does transformation or redemption look like when one takes the earth seriously? Most theologians, like Lonergan, did not have the ecological crisis in mind when they constructed their theologies. While modern theology, in general, focused virtually exclusively on the human subject, a few, like Lonergan, constructed systems that are eminently open to include new issues and problems as they arise.[40]

In 1996, Dieter T. Hessel concluded that by and large the environmental teachings and practices of most Christian churches in the United States were concerned with the impact of the ecological crisis on human communities. It was part of a new social gospel movement. Hessel noted that comparing the new social gospel with the social gospel at the turn of the twentieth century revealed a few characteristic movements beyond a totally anthropocentric concern. These movements included such explicit concerns as animal welfare and energy conservation, but most efforts to date had been a form of benign stewardship that still viewed the earth as instrumental in human well-being and human-divine relations.[41] So the ecojustice movement at the time, while benefiting both earth and humans, had as its primary focus justice for poor communities.

A few religious ecologists (Thomas Berry was one) resisted the integration of ecology concerns into social justice concerns as well as the biblical notion of stewardship of the earth precisely because of this tendency to lose focus on the nonhuman world almost entirely. However all, including Berry, saw the problem as being a primarily human one in that it was humans who caused the problem and if anything was to change it was humans who had to change. Clearly, this meant more than individual conversion, traditionally understood, as the cultural and societal decline that had been complicit in the crisis also needed renovation. The degree of renovation advocated varied across a spectrum from radical change to something more cosmetic, such as adding earth care to the list of social concerns. In terms of breaking the longer cycle of decline as the ecotheologians themselves described it, the gradual real move to integration of ecological and justice issues, the understanding of a world global system in which both ecological and justice issues are indicative of a suffering world, is what the authors of this text see as holding the most promise. While many different theologies can lead to greater responsibility for the earth, the most comprehensive views seem to respond best to the attentiveness, understanding, responsibility, and commitment advocated by Lonergan as necessary for redemptive transformation.

When viewed as a community of texts, the texts that have been produced by Christian ecotheologians call for a massive revision of an entire way of life, from corporate systems and governments to theology and church organization, to individual spirituality. In their foreword to the series *Religions of the World and Ecology*, Mary Evelyn Tucker and John Grim identify the overall aims of the conferences that led to the series publication. One of those captures best the kind of transformation that these texts call for: "To articulate in clear and moving terms *a desirable mode of human presence with the earth*."[42] Contributors to that conference and the resulting texts represent the broad spectrum of elements of transformation. Of course, it is not within human capacity to judge with any certainty that individuals or communities have achieved inner transformation, described by Lonergan as the love of God flooding one's heart. However, the calls for soul-searching examinations of life styles, social systems, and traditions indicate a radical and authentic transformation that values creation as God's and seeks to establish right relations among God, humans, and the created order.

Elizabeth Johnson, Sallie McFague, and Mark Wallace seek to recover the values and teachings within the Christian tradition that might prove helpful in giving meaning to a new ecological praxis.[43] Johnson reclaims from the historical tradition forgotten or ignored creation teachings; McFague rereads understandings of Christology to focus on the embodiment of God and intimate relationship of God and creation; Wallace suggests new interpretations of the presence of the Holy Spirit in an ecologically suffering world. These are just a few of the ecotheologians delving into strongly held and sometimes hardened teachings to give new meaning to their tradition and to lead others to an enlivened faith for the sake of the whole world (human and otherwise).

Respondents to these reconstructions point out the necessity of atoning for the abusive way in which such doctrinal beliefs have been lived in the past and of giving the renewed doctrines power in practice. By reconsidering creation anew, Gordon Kaufman suggests, we can find a better niche, a new way of living within the ecological order.[44] We need to consider the repressive force buoyed up by the notions of the universal and cosmic Christ associated with the power of the Roman state as Christians became entwined with it, according to Kwok Pui-lan.[45] Eleanor Rae adds to Wallace's understanding of the Spirit, the biblical association of Wisdom with the Holy Spirit. "The language of Wisdom is the language found in the public places, the places where people are gathered: the streets, the plaza, by the city gate (Prov. 1:20–21).... It is more necessary today that the voice of Wisdom once again be heard in the public places."[46]

Some ecotheologians attempt to bring traditional Christian values, motivations, and practices to bear on specific issues, indicating concrete ways in which contemporary Christians can live traditional values. David G. Hallman, long-time worker on ecology issues with the United Church of Canada and later also with the World Council of Churches, sees "climate change as a metaphor for the fractured relationship between human societies and God's creation."[47] He calls for a moving away, by individuals and larger social systems, from life-styles that "are defined by their material wealth." This would allow us "to move toward life-styles that are defined by their attention to relationships." Justice as understood by the Christian tradition demands this change.[48] Larry Rasmussen proposes "sustainable communities" based on an extension of Christian values and way of life to include nonhuman nature. "The constructive ecclesial task," he argues, "is both long-term and immediate. The pressing immediate task is to address, one by one, the downside effects in city neighbourhoods of the global economy's impacts and to try to find, in practice and policy, those incremental changes that slowly build toward sustainable communities.... The long-term task, succinctly put, is the conversion of Christianity to the Earth."[49]

While all of the ecotheologians considered here recognize that the ecological crisis demands a new way of being spiritual, some focus on this personal transformation. Speaking out of the apophatic tradition of Eastern Orthodox Christianity, John Chryssavgis comments,

> I believe that the apophatic dimension is wonderfully, indeed "naturally," fostered in creation. The breadth and beauty of this earth is a reflection of the boundlessness and splendour of divine grace; and our respect for the environment results in a parallel allowance for the surprising abyss of God. Our admiration for creation reflects our adoration of the absolute, a vocation to the beyond, an invitation to transfiguration.[50]

Similarly in a different context, Douglas Burton-Christie connects prayer with attentiveness to the voices of the earth as heard especially in the words of nature poets. He cites an excerpt from the poem by Mary Oliver, "Have you ever tried to enter the long black branches?"

> Only last week I went out among the thorns and said to the
>   wild roses,
> deny me not,
> but suffer my devotion.[51]

Then Burton-Christie asks: "Is this not the language of prayer, the most intimate and probing of all languages? . . . Such language [direct address of the wild roses and the bold command] reflects a reverential respect, even awe, a sense that one is approaching the numinous other. So, too, does the language of devotion. To devote oneself is to vow oneself, consecrate oneself to the Holy One."[52] He suggests that the broken relationship with the earth and the cosmos is the reason for our inability to hear the Word in creation, and to pray in this deep renewed sense. And furthermore, the rekindling of a relationship with the concrete places of the earth might require this renewal of intimacy. "It may also be necessary," he concludes, "to the long-term survival of those places."[53]

In these few representative texts of ecotheology, many levels of transformation are addressed. The reconstruction of worldviews and the redirection of praxis and spiritual life are all required in meeting the long cycle of decline that contributed to the present state of ecological systems. Theologically, the reimagining in Christian terms of everything, from notions of the nature of the Divine to the content of daily prayer and practice, are seen to have an impact on the present social imaginary toward a more sustainable manner of human life.

5

SCIENCE AND ECOLOGY

This chapter examines the use of science by ecotheologians. We have described the ecotheological texts as engaged. These texts take into account aspects of the world that relate to the ecological crisis and seek to promote meaningful Christian practice in response. Since their concern is with the natural world, physical science is one of the more consistent conversation partners. We contend that the force of the texts does not lie primarily in how accurately scientific theories and other content is reported. Rather it lies in a critique of the scientific process itself. The focus is on the agendas and ideologies that Western science like theology have carried and how science too has been complicit in the ecological crisis.

Recent works by theologians who are also expert in some field of science give convincing critiques of some uses of science in ecotheology and indicate how a more accurate use of science can enable improved moral practice. These works are generally relatively recent. They respond in part to the increasing public awareness of climate change and the need for action. As the authors of *Ecospirit* so clearly ask: What has to happen to convert ecological awareness to ecological action on a large scale?[1] Biological science in particular (but not exclusively) gives insight into the behavior of humans as products of evolution. This is helpful in understanding and motivating ethical practice on behalf of the natural world. In terms of the social imaginary, these texts make available a set of tools that can enable a change of practice and a new ecological turn in the social imaginary. In Taylor's words, the social imaginary is "not a set of ideas; rather it is what enables, through making sense of, the practices of a society."[2] Predominantly, the ecotheological texts we are examining aim to initiate new ecologically responsible practices and to make sense of these within a religious tradition.

When these texts engage the best scientific understanding of humans and their evolution they support responsible ecological practices.

Not all ecotheologians use science in the same manner or with exactly the same intent. We have identified a number of strands of the interplay of theology and science. These strands represent differences in emphasis and are best understood as positions on a continuum. On one end of the continuum are those texts that see science (in some sense) as providing a cosmology that promotes the radical change of consciousness required to meet the ecological crisis. On the other end are those texts for which science fulfills more pragmatic functions. They educate and ground ecotheology so that it is not simply dismissible by physical and human sciences. They describe human behavior in a way that grounds a concrete and effective moral ethic. They provide scientific data that help make a case for ecological action, most often on issues of ecojustice. Between those two poles (science for a change of consciousness and science as a basis for more direct just action) lie various permutations and mixtures. There are those who emphasize ecosystems to support an organic sense of community and kinship among living species. There are also those for whom science provides a right understanding of the natural world or of human behavior on which to base a more sophisticated stewardship and moral ethic. Outside the continuum are those whose theology responds to the ecological crisis without any analytic use of science—these simply accept scientific descriptions of ecological ills and their prescriptions for remedies.

SCIENCE AS COSMOLOGY

The most prominent use of science in the construction of a cosmology to ground an ecological ethic is Thomas Berry's and Brian Swimme's *Universe Story*. While this text is not ecotheology in itself, it calls for a grounding of all religious renewal within a proposed new cosmology. The new cosmology is a construction of contemporary understandings of the evolution of the universe and the earth, from the Big Bang to the emergence of human cultures. It employs traditional religious symbols and myths to create a meaningful account. The evolutionary process becomes a history of the universe as well as the spiritual journey of the universe. This construction takes seriously the evolutionary findings and theories of the physical sciences and the history of cultures, but it also rests on a critique of Western science as overly materialistic. Such a use of science typically views science as a cultural artifact, not as the all-encompassing explanation of reality. Following critiques such as those of Ilya Prigogine, Berry holds that the science that developed in the West inherited ideas already encompassed in the societies of Europe. Ideas such as the independence of the created order from its creator, the lawfulness of creation, and the separation of soul and body were inherited from the Bible and its history of interpretation. They became basic assumptions of Western science, and hence science itself is now seen as complicit in ecological degradation. Like theology, science carries ideologies. It is not, as was traditionally held, value-free. However, scientific "facts" cannot be explained exclusively by epistemology. Berry and Swimme and others who create cosmologies on the basis of scientific findings are neither naive realists nor postmodern constructionists. They do, however, promote a holistic view of the universe. Spirit and matter are mutually implicated in the evolution of the cosmos. No one part of the universe is privileged. The human species is an integral aspect of the universe. For example, Berry describes the human as that being in which the universe becomes conscious of itself. In other words, the human functions within the universe for the good of all beings. Like all other beings, humans have a particular role that works along side all other roles in the creation and sustenance of the whole and its component parts. However, without all other aspects of the universe there could be no human life as we know it, just as we would not have this particular universe without the presence of humans along with the other aspects. The human role is not a higher or more privileged role, but it is a role that defines the human species.

What gives science its power for Berry is primarily its mythic quality. Myth has traditionally functioned within cultures to bring intelligibility and value to human affairs. Science assumes a similar function within modern societies. It lends credibility to human endeavors of all kinds—politics, economic decision making, and health care, for instance—and certainly to environmental policies and actions. For any ecotheology to have validity, then, it must grasp an adequate account of science. The scientific account of the evolution of the universe must be the context within which all human affairs are conducted. In Berry's words: "All human professions, institutions, and activities must be integral with the earth as the primary self-nourishing, self-governing, and self-fulfilling community."[3] This is the functional edge of the new cosmology—a story of the universe, in all its unity and diversity that has the power to evoke a new ecological consciousness and sound ecological practice.

As a cultural historian, Berry is well versed in the power of ritual and symbol in the formation of human consciousness and human cultures. The whole corpus of his work reveals a sophisticated vision of the practical and functional way stories and myths change behaviors over time. In this there is hope for the ecological crisis: the power of science within Western culture, at least, to redirect the human psyche toward a greater valuing of the natural world. According to this view it is not only the empirical facts, the laws and numbers, as recounted by science that are considered reality. Radiance and beauty, risk and violence are as much a part of the universe as the empirical descriptions of science. As Alfred North Whitehead observed, one can have all the empirical knowledge but fail to be inspired. In his words: "When you understand all about the sun and all about the atmosphere and all about the rotations of the earth, you may still miss the radiance of the sunset."[4] For Berry and others who use science as an all-embracing cosmology, both knowledge and inspiration are needed to address the ecological crisis adequately. Those who possess both as an integrated story are entering a new ecological age in which ecological meanings and values ground a new ecological way of life.

A number of Christian ecotheological texts that focus more intently on the renewal of Christian theology reflect views similar to Berry's. In 1972, William Pollard summarized the prevailing account of the evolution of the universe, including the historical development of humankind, to highlight the tragedy of our ecological predicament and to inspire a Christian response to its challenge.[5] Several articles in the volume in which Pollard's essay appears also take account of the evolutionary nature of the universe.

A more developed and significant example of this approach is Rosemary Radford Ruether's *Gaia and God*. Ruether announces her intention with regard to use of "the scientific story of creation" as follows: "In this chapter, I wish to take a few steps in the direction of asking how the story of creation that has emerged from earth history might also function as creation 'myth' in the classic sense of mandates for ethical relationships."[6] It is necessary that "scientist-poets" retell the story of the cosmos and earth's history in a way that replaces the dualism between knowledge and wonder, "in a way that can call us to wonder, to reverence for life, and to the vision of humanity living in community with all its sister and brother beings."[7] Ruether challenges engrained perceptions of the natural world as "red in tooth and claw" and "survival of the fittest." These continue to feed the aggressive, competitive, and violent tendencies of Western societies.[8] To counter this, she emphasizes the interdependence and co-evolution of beings within the universe. She also critiques negative Western perceptions of death and decay and calls for a clearer distinction between evil and finitude in Christian theology. Evolutionary theory and the intimate relationship of humans to the rest of the world support her critique. In Darwin and later evolutionary theory, death has a crucial role in the furthering of life. In Ruether's view, Christian teachings that have understood death only as a result of sin ignore natural processes. At the human level they promote more violence by projecting blame on exploited victims. Humankind must accept its place as a finite species within a created world characterized by interdependency.[9]

Gaia and God systematically addresses ideology and domination (predominant themes throughout Ruether's work) and their effects on both the natural and human social world and places these concerns in a cosmological context. She reminds the reader of the collapse in scientific understanding of the distinction between energy and matter at the subatomic level. Reflecting on the ways in which other animals experience life, she concludes that human consciousness must be understood as existing not only in continuity with all other life, but also as set in complex and fragile contexts of body and ultimately cosmos.[10] Furthermore, as is characteristic of the use of science as cosmology, Ruether calls for a renewal of human consciousness, for an "ecological culture" that would convert human consciousness "to the earth." This conversion ought to be expressed in community ritual and action.[11] Ruether's texts, in particular those that enable the voices of women from the majority world, make other uses of science, as will be noted below. However, her own indications of how theological renewal might proceed are set within the cosmological vision she describes in *Gaia and God*.

Anne Primavesi, too, talks about Gaia in constructing her case for a serious dialogue between science and religion to undergird and foster ecological responsibility. In her preface to *Sacred Gaia*, Primavesi argues that any significant contribution by theology to environmental discussion must necessarily show familiarity not only with scientific environmental language, but also with the understanding that "ecosystems to which we all belong interconnect within a greater whole."[12] Her work was written in conversation with James Lovelock's holistic co-evolutionary theory, for which he uses the metaphor of Gaia. According to Lovelock (among others), organisms and their physical and chemical environments co-evolved. As each responded to the other, organisms have helped create the situations in which they thrive. Similarly, physical and chemical components of the earth have given rise to organisms that participated in the continued change and adaptation of these components.

On the larger end of the spectrum, the whole cosmos itself is understood to be a self-creative or autopoietic entity—self-making and self-organizing—but not self-interpreting.[13] Everything from the metabolic processes of our own bodies to the large time and space scale of cosmic emergence is a single evolutionary process.[14] For Primavesi, this picture of the cosmos, framed by the Darwinian resituation of the human in the universe, means that theology itself must be an earth science. Theology must readdress issues such as our self-perception as humans, our sense of justice to the whole earth community, and the consequences of our "God-concepts." *Sacred Gaia* addresses these issues systematically, all under the overriding concern for the devastating presence of the human species on the planet earth.[15]

Primavesi shows why contemporary theories about the origin and continued emergence of the universe ought to affect theology. In doing so, she carefully attends to the present debates about the scientific epistemology and postmodern awareness of the role of human language in our perceptions of reality. Difficulties in distinguishing life from nonlife, for example, or one species from another, are common in science. Central to her argument is the difficulty of de-"anthropocentrizing" evolutionary explanations using human language, in particular theological language. That's why one must exercise care (e.g., attention to particularity of time and place) in the formulation of theological meanings.[16]

In *Sacred Gaia,* science is used to express and elaborate an understanding of the whole—the universe—in order that the concept of sacred, when applied to Gaia, has a concrete basis. Primavesi holds that the term sacred cannot be applied merely to particular beings or locations; it is the whole. This wholeness of the sacred disallows hierarchical binaries where the term denoting "them" (of whatever kind) is devalued. In her words, "A major task for theology as an earth science is to resist any process or tendency toward such devaluation by stressing connectedness, diversity, and sacredness of all beings."[17] God must be allowed to be "God of the whole earth system."[18]

Gaia's Gift: Earth, Ourselves and God after Copernicus follows Primavesi's *Sacred Gaia,*[19] focusing more specifically on our self-perception as humans and the need to confront the anthropocentrism that sets the more privileged among us above all creation, made in the image of God. Post-Copernican perceptions, in her view, have not converted humans to an earth-centered view. This is significant because all scientific perceptions are based on human observation and human judgments are earth-centered. Human perceptions of our place on earth may change, but in Primavesi's well-argued view, this does not change human self-perception as master and user of the earth. However, Primavesi contends that present cosmological and ecological science that takes into account the current ecological crisis is forcing us to perceive our roles differently.

"Neither the role of the living beings nor the role of ourselves has changed. However, our perceptions of both roles—and in particular of their interconnectedness—is gradually changing as we experience, observe and deepen our understanding of their interactive effects."[20]

For Primavesi, it is the mythic and metaphorical elements of the Gaia theory, the emotional connection to earth, and scientific knowledge that bring about this change. She argues that rationality and myth are not opposites but are both dimensions of human knowing: "myth is a way of teaching us the kind of relationship with nature that supports rather than destroys a community's life."[21] Primavesi sees that scientific accounts on their own lack the gift-giving dimensions of Gaia and the call for an exchange of giftedness by humans. In order to celebrate and live as if all creation is to be enjoyed in common, humans must see the scientific account of the cosmos in its mythic and symbolic as well as empirical dimensions. Primavesi uses the work of Ernst Cassirer (among others), to argue that "when language is viewed symbolically, it is not a product but an activity"[22]—part of the network within but also beyond a culture.

Process theologians use science as cosmology in a similar but distinctive way. Their work is primarily inspired by Alfred North Whitehead, but those whose theology is also ecotheology sometimes blend concrete scientific theory with Whiteheadian philosophy. Having already understood the universe and God's relationship to it as process, it was not a large step for some of the process theologians to incorporate evolutionary theory on behalf of the ecological crisis. The most outspoken and prolific contributor to this form of ecotheology is John Cobb Jr.

Cobb was drawn to Whitehead through the influence of his teacher, Whiteheadian philosopher Charles Hartshorne. Cobb's own account of the move he made from a process theology based on Whitehead to a process ecotheology is illustrative of this use of evolutionary cosmology. In 1965, he published *A Christian Natural Theology*, in which he admits, "I wrote about God and human beings based on Whitehead's philosophy and said almost nothing about the rest of the natural world."[23] As related in chapter 3 above, Cobb reports that he began to see the ecological implications of Whitehead's philosophy. Within his now ecologically sensitive rereadings of Whitehead's philosophy, Cobb became aware of the relevance to the natural "other-than-human" world of the value of all experience to the Consequent Nature of God. He became convinced that attributing value only to the human world was "totally erroneous."[24] Key insights include awareness that all beings from the smallest sub-atomic particle to the largest planetary system, across time, are capable of an experience intrinsic to themselves; all contribute to God's experience and nature. This approach, similar also to that of Charles Birch, highly values the entire created order and views humans as intimately at home in the cosmos.[25] Process thinkers are often criticized by ecologists for the hierarchy of value, giving more value to some entities than to others based on complexity. Humans therefore are attributed the highest value because of their high capacity to integrate greater diversity.[26]

Yet process theologians such as Cobb find in their synthesis a meaningful valuing of science, the whole cosmos and all its constituents, and human responsibility for the world. Jay McDaniel, another well-known process thinker and ecotheologian contests this interpretation, "Whitehead seeks to overcome the separation of facts and value that is so common in modern thinking. . . . Whitehead believes that there is also a certain kind of value in the world even apart from human projections: namely, that which other living beings have in themselves as they struggle to survive with satisfaction. . . . This kind of value was part of the earth long before humans evolved."[27]

John Haught has consistently engaged science in his version of process ecotheology. Haught's major preoccupation has been the refutation of what he calls "cosmic pessimism"—the naturalist view that all is explainable within the material laws of nature and in the end the whole cosmic process is meaningless.[28] Richard Dawkins, Daniel Dannett, and E. O. Wilson are well-known advocates for the naturalist position. Influenced by the work of Bernard Lonergan as well as Whitehead, Haught addresses the purposefulness of the universe not by the argument from design but by the future-oriented hopefulness he reads in the synthesis of contemporary science and religious traditions. In his most recent systematic treatment of purpose in relation to scientific materialism, he argues that the cognitional structure that is at work in the minds of naturalist scientists themselves cannot be explained by their own theories.

Following Lonergan, Haught lays out a threefold cognitional structure and imperatives of the desire to know: experience (be attentive), understanding (be intelligent), and judgment (be critical). He challenges the naturalists to account for the evolution of this combination of imperatives within their materialist evolutionary theories. "Most scientific naturalists," Haught argues, "adhere passionately to certain values whose power to motivate would be immediately deflated if their own naturalistic accounts of the origin of virtue were applied to these values."[29] Haught argues that creation contains the sources of all that exists, not on the basis of science alone but also the presence of a self-abdicating God who grounds reality in all its emergent capacities. Only when science and religion take serious account of each other can one achieve an adequate account of creation.

Process theologians critique and respond to the separation of scientific and other perceptions of the world that they claim to be complicit in the ecological crisis. They share this view with other ecotheologians—Berry, Ruether, and Primavesi, for example. All claim that a more organic view of the whole better represents reality and is more powerful mythically and symbolically in promoting sound ecological practice.

SCIENCE AND THE EARTH COMMUNITY

Other ecotheology texts focus more on planet earth and rely primarily on the science of ecology. Sallie McFague has made a notable contribution to the literature of ecotheology using this approach. A dominant theme of her work has been the reforming of theological doctrine in terms of metaphors based on ecological science. If one takes the ecological nature of the earth seriously as a context for theology, then the most appropriate way to speak of God is as the body of all that exists.[30] She proposes this model of God not as an exclusive model but one that takes seriously the organic world and the science that describes it, while at the same time attending to an ecologically damaged world. In her words, "The model of the world or universe as God's body is, I will argue, in keeping with the view of reality coming to us from contemporary science. . . . As a theologian, I am concerned with the broad parameters of the contemporary scientific picture of reality not only because the credibility of faith depends upon that connection, but also because the contemporary view tells us about our world in ways we desperately need to hear and heed."[31] McFague proposes an incarnational theology, taking seriously the embodiment of God in Jesus Christ, in which human beings are "of the earth" primarily; the earth is a home to be shared with all created beings.

Like the other ecotheologians discussed in this chapter, McFague relies on the contemporary scientific creation story and interprets it to privilege embodiment as context for all human knowing and doing.[32] The creation story is characterized by (1) a vast range of time, beside which the notion of God's concern merely with humans "dwindles"; (2) history-creation that is a moving, dynamic process; (3) the organic character of radical interrelatedness and interdependence; (4) a complex and multileveled universe within which subjectivity is always present and emergent; and (5) an accessibility to all cultures, which are free to remythologize the story within their own contexts.³³ In this characterization of the scientific account of evolution, McFague grounds reinterpretations of traditional theological doctrines, in particular of the nature of God, but also of sin and redemption. Even *Abundant Life*, which focuses on the implications of economic realities for theology, does so within an ecological context. Economics also must be ecological, she claims. It is, "first of all, a vision of how human beings *ought to live* on planet earth in the light of the perceived reality of *where and how we live*. We live, in, with and from the earth. This story of who we are is based on postmodern science—not on classical economics."³⁴ McFague shares with other ecotheologians a concern that her reconstructed theology promotes a critical view of the world and an ecological praxis. The human relationship to earth as described by science has a moral edge. Humans have a place in the created world; that place ought to be inhabited in a way that is just for all other created beings.

The notions of body and the interrelatedness of all creatures is now common among ecofeminist theologians who follow the thought of McFague and Ruether in using scientific accounts of evolution and ecology as a context for their theologies. Ivone Gebara, for instance, speaks of the individual human body as a Cosmic Body because of the intimate connections of humans to the entire universe. She also calls our "larger self" the Sacred Body of the Cosmos.³⁵

This book does not discuss works of Christian eco-ethicists in detail. However, it is worth observing that those who deal directly with bases for ecological ethics seek to expand their ethical frameworks to account for the latest scientific descriptions of the earth (and often the entire universe). In *Earth Community, Earth Ethics*, Larry Rasmussen argues that scientific data are as important for ethics as are theology and philosophy. He writes, "An ethic will not command allegiance if there is no resonance with the universe and the world as we experience, know and describe them."³⁶ Rasmussen moves from the consideration of the earth as a whole to consideration of human interactions with creation within particular communities. In a volume Rasmussen edited with Dieter Hessel, various authors present examples of earth-honoring Christian communities. These instances are set within the context of earth as the primary habitat of humans.³⁷ Similarly, James A. Nash grounds his moral theology on the relationality of God not only to humans but to the cosmos and the earth. Kinship with all creatures on earth and the human responsibility that derives from that kinship is for Nash the basis of Christian morality.³⁸ The title of Denis Edwards's book, *Ecology at the Heart of Faith: The Change of Heart that Leads to a New Way of Living on Earth*, also reflects the turn of ecotheologians to the physical sciences as a context for theology.[39]

There have also been critiques of the manner in which most ecotheologians incorporate science into their renewed theologies. Lisa H. Sideris wrote a book-length evaluation of ecotheologians' incorporation of biology into their work. Sideris argues that in spite of the apparent turn to ecology by ecotheologians, most do not take seriously evolutionary theory understood as natural selection. While ecotheologians do purport to use an ecological model, that model does not correspond to a Darwinian or neo-Darwinian view. Using Ruether, McFague, and Jurgen Moltmann as examples, Sideris notes that they rely heavily (in some cases exclusively) on the notions of nature as interdependent, interrelated, and harmonious. From this premise they argue that humans are to extend an ethic of love and community interdependence to all of creation on the basis of how creation actually is. Sideris claims that not enough attention, and sometimes no attention, is paid to the suffering that is part and parcel of evolution. When the suffering of nature is discussed it is understood as oppression of nature by humans, a category taken from liberation theology (and thus a human-to-human category). How, she asks, is one to understand the healing of nature especially in the work of McFague where, for example, the emphasis is on the healing of bodies? Natural selection cannot operate on the basis of healing of individual bodies. Is it in the interest of nature to heal individual wild animals out of human compassion? Does this respect the integrity of wild animals and ecosystems? What about the predator-prey relationship? Sideris claims that these questions are not adequately addressed by most ecotheologians.[40] She concludes her critique of ecofeminist views of nature as follows: "The community model of nature (or a single community of humans and nature) is promoted as a key ecological insight in these arguments, and, once again, evolutionary processes are either omitted or reinterpreted in light of this ideal."[41]

Sideris also studies process theologians (specifically Birch and Cobb, who do admit the strand of suffering within evolution). However, she critiques the notion of richness of experience as a basis for how one would ethically treat different creatures. As she points out, within ecosystems it is the role each creature plays in the system that is important, not the richness of experience extrapolated as distance from (or closeness to) human capacity to experience. To illustrate her argument, Sideris raises the issue of the protection of endangered species. "Freshwater mussels, for instance, are currently one of the most endangered species in the United States," she observes. These animals are vital to many ecosystems, but the general public seems quite unconcerned. The same can be argued for insects and even microscopic animals and plants. A valuation of species based on richness of experience as described by Cobb and Birch does not offer an adequate ethic for protecting nature on its own terms.[42] The valuation of other-than-human beings by process theology, then, remains anthropocentric. Further, we might add, only adequate legislation, vigorously enforced, can protect endangered species.

While the case can be made that nature "as it really is" (Sideris's term)[43] is not fully knowable and therefore the nature to which we respond is always subject to interpretation, it is critical that ecotheology use the best and most generally accepted science of nature.[44] Sideris's proposal is for a more nuanced reading of nature (including human nature) as subject to natural selection. Critical examination of the human response to suffering in nature must be tempered by the best science on diversity and difference, interdependence and interrelatedness in nature. Compassion may well be appropriate in the case of domestic animals, for example, but may be problematic in terms of individual animals in the wild. Countering Sideris's argument, some ecotheologians (e.g., McFague and Ruether) may point out that their intervention is focused on change in larger socioeconomic structures. Such structures have negative results for all of nature, especially for the poor and for nonhuman creatures. Larger socioeconomic structures are much more systematically confronted by these ecotheologians than by Sideris herself. The data of the social sciences and economics regarding the status of the poor and of women, especially in the majority world, and of the impact of the global market system are taken seriously in the works of Ruether and McFague and others discussed in chapter 6.

It is not the purpose of this brief presentation either to support or counter Sideris's critique of many ecotheologians (not all of whom are presented here) or to do justice to all her scientific and theological arguments. The purpose is rather to indicate that the conversation between the natural sciences and theology for the sake of responsible ecological practice is a dynamic one. Furthermore, the texts of the early ecotheologians who challenged the agendas and ideologies of both science and theology have resulted in an engagement both in ideas and in attempts to reform attitudes and actions. Critiques such as Sideris's continue the engagement precisely because effective ecological ethics is a shared goal. In Sideris's words: "My intention is not so much to test their theological claims with exclusive reference to scientific data or secular perspectives but rather to illustrate how a scientifically inaccurate understanding of nature perpetuates ethical and theological imperatives that are inappropriate."[45] Action that makes a difference in the way humans treat the rest of creation remains the horizon for the conversation in ecotheology between theology and science.

Whereas Sideris calls for an ecological ethic that takes seriously primarily other-than-human nature, biologist and theologian Carolyn King examines the theological and ethical implications of humans as products of natural selection. While she is aware of the roles of humanly constructed systems, such as economics, in the present ecological crisis, King makes her specific contribution in an analysis of the biological roots of human behavior: "Here, I concentrate on the *biological* reasons why the challenge to humanity, to learn to live within its limits, is at the same time so obviously necessary and so impossibly difficult."[46]

King does not claim that innate tendencies that arise from natural selection are the only determinant of responses to the ecological crisis. Humans are also coded for cultural production and can and ought to operate within a value system that transcends our naturally selected tendencies. Nevertheless, King contends that Christian teaching regarding the response to the ecological crisis must begin with a significant understanding of how and why it was that humans came to be the way they are. The theory of natural selection, in Darwin and in its later development in genetics, for example, is the essential starting point for any analysis of human behavior. King's training and experience as a biologist gives her a fluency in scientific theory and in a wide range of animal behavior. Joined with her sensitivity to what she calls the "habitat of grace," her text provides a lucid account of natural selection that is understandable by most lay readers.

Like Sideris, King decries any romantic notion of animal behavior as expressive of compassion as humans understand the concept. Most if not all cooperative behavior that could be misinterpreted as compassion in a human sense can be explained by the innate drive of organisms to hand on their own genes. Female foxes will often cooperate in order to rear younger cubs in the pack rather than reproduce their own when cooperation better ensures the continuation of their genes. On the other hand, a dominant male lion will devour the young of another male so that the mother will carry his young and so propagate his genes. However, this is not an indication that foxes are more compassionate than lions.[47]

Similarly, humans *naturally* will act in the interest of survival for the sake of handing on their genes. "Look out for number one" is a genetically inherited instinct. Sociability works because self-restraint for the sake of the group also benefits the individuals within the group. In order for this to work, the group must be small enough that the benefit to all is obvious and felt. Work in the area of game theory supports this observation. The classic case of the Prisoners' Dilemma is an example that helps explain the tragedy of the commons (the tendency of some individuals to grab more than their share even if it destroys a shared resource).[48] The Prisoners' Dilemma refers to a fable of two burglars who are questioned separately by the police. Each of the prisoners has to decide whether or not to confess, inform on the other, or remain silent. Having considered all of the options (e.g., burglar A informs on the other, while B remains silent but goes to prison) the conclusion from each burglar's viewpoint is that the safest option is to betray the other. Applied to the common use of goods, this translates into getting the most of a resource (pastureland, water, fish, and so on) for oneself, despite the fact that restraint on the part of all would in the end benefit all the most. However, it is unlikely that under normal circumstances any beneficiary would practice restraint without a guarantee that all were practicing similar restraint.[49]

King does not propose that we should understand natural selection or game theory because nature is an appropriate model for ecological responsibility. Rather, a clear understanding of natural selection makes us aware that in most cases we are not inclined by nature to be ecologically responsible. The source of motivation has to come from somewhere else because, as King explains: "On a wider scale, environmentalism in general is the same issue—it is a form of the Prisoner's Dilemma game involving many players, and the problem is how to prevent egoists producing pollution, waste and exhausted resources at the expense of more considerate citizens."[50] Yet, natural selection produces not only negative instincts; instances of reciprocity and kinship are well known in nature and probably form a basis for human compassion. However, in the current ecological crisis, such bases are not sufficient to ground global responsibility. Here, according to King, is where religions can play a significant role.[51] Together with several scholars to whom she refers, King holds that the true role of religion is not merely to recognize the significance of science and support its efforts in the ecological crisis; rather it is to understand who we are as humans products of natural selection and subject also to the influence of culture. In the end, as we are contending in this book, it is the values and way of life that are imagined and lived within human societies that will be critical in whether or not we adequately confront the crisis.[52] The implication of King's work for Christian churches is further examined in chapter 7.

A relatively new concern for ecotheologians is the field of bioethics and its implications for the ecology of the planet. Celia Deane-Drummond is among the first Christian theologians to address this issue from an ecotheological perspective.[53] Deane-Drummond acknowledges the importance of biology in the construction of human ethical behavior. However, she cautions that the "ethical commensurability" of humans and nonhumans is not such that biology or ecology is a helpful basis for an eco-ethics dealing with biogenetic issues.[54] She argues instead for a virtue ethics based on a divine-initiated cosmic covenant that sees humans as participants in the whole of life. She finds the evidence for this covenant in biblical accounts of creation including the eschatology of Isaiah. The whole of life is a gift of God. Virtues such as friendship, compassion, charity and justice enable humans to respect difference but at the same time give appropriate care and respect to all creatures. Wisdom and prudence understood within a life of participation in God's covenant with the whole of the cosmos enable human judgments governing how humans deal with bioethical issues such as cloning and transgenics.[55]

Having discussed the various contemporary controversies around biogenetic research and applications, Deane-Drummond concludes that many of the stated aims as well as practices of biogenetics transgress "an ethic of wisdom in the context of love and friendship."[56] The cosmic covenant initiated by God demands prudent care. Humans are encouraged, indeed required, to be attentive to their fellow creatures' well-being not because of the nature of the biological order, but because of the nature of the relationship among creatures established by God. For Deane-Drummond, the science of ecology or of biology enlightens ecotheology not by providing a paradigm for human community or behavior, but by elucidating the complex variety of interrelationships by which humans participate in the cosmos. This is informative and helpful knowledge that enables a wise and loving eco-ethics.

A number of theologians who are seriously engaged with ecological issues use science in much more practical and directly functional ways. For some of those, such as Calvin DeWitt, theology calls for ecological responsibility or stewardship. For them the basis for this call lies within the biblical and church traditions themselves. The role of science is to provide the knowledge necessary for correct action. In other words, science becomes the means to fulfill a Christian vocation. This form of ecotheology is sometimes called evangelical environmentalism, and is the term characteristic of evangelical Christians. Biblical stewardship is a broader term also used by evangelical Christians, usually quite interchangeably with evangelical environmentalism. DeWitt, a pioneer and figure of enormous stature in this field, resists the conflation of the two terms (biblical stewardship and evangelical environmentalism). As he says, evangelical environmentalism "is not a term I really welcome, largely because evangelicalism works to see the creation whole; there is no 'us' versus the environment. Rather, human beings are part and parcel of the creation even as they are made in the image of God, and the creative system is not separable into us and everything else."[57]

DeWitt is founder of and active participant in a group of scientists teaching in more than 60 colleges and universities who identify themselves as evangelical Christians, and who meet yearly for serious study of environmental issues. Representatives of this group (scientists, theologians, policy makers, and others) have been meeting as the Au Sable Forum for decades to study issues of science, research, and teaching in the context of serious Bible fellowship. Their motivation is deeply connected to their Christian beliefs and their desire to make the world right, following Jesus. "The purpose of Bible study extends beyond edification to discovering Biblical teachings and applying these to personal lives, society, and the rest of creation."[58] The interplay of science with this theological stance finds its basis in the further theological understanding of how one knows God, not only through the Bible, the book of God's words, but also through creation, the book of God's works. In this strand of science/theology interplay, science enables wise interaction with the rest of creation. In the light of the ecological crisis, science provides the knowledge for the right practices involved in promoting the biblical injunctions to be stewards of God's world.

The Centre for Studies in Religion and Society, founded by and led for most of its existence by Harold Coward, has a multireligious base (that is to say, not exclusively Christian), yet operates in a similar fashion. The vision behind this center is to engage a group of scholars, policy makers, and others in bringing together religion and issues of import to society. This center is not primarily concerned with a reformulation of existing theologies. Its goal is to foster the work of scientists, religious scholars, and others in confronting major world issues. Ecological topics have been well represented in the meetings and publications of the center. As in the Au Sable Forum, it is the scientists and the religious scholars themselves who are engaged and not only texts they have produced.[59]

Part of the controversy and critique over the use of science among ecotheologians involves various interpretations of the earth "as it really is." It can be amply demonstrated in the work of all of those theologians presented in this chapter that none espouses a naive realism that claims to know the earth "as it really is." However, all do maintain that there is a real earth beyond human epistemology and that it can be reasonably known, for example, through modern scientific methods. They hold a critical realist position by which they view human knowledge as being at least in part a product of the various ideologies, attitudes, agendas, availability of data, and so on that constitute the culture, society, and individual location of the knower. Certainly such a view of human knowledge is acknowledged in the texts of most ecotheologians.

Yet another perspective on paying attention to the earth as it is and furthering a conversation between science and ecotheology is that of Brad Allenby. Allenby works in a multidisciplinary world: he is a vice-president at AT&T responsible for environment, health, and safety, an adjunct professor in the School of International and Public Affairs at Columbia University as well as at the University of Virginia Engineering School, and lectures regularly at Princeton Theological Seminary.[60] Allenby calls on all environmentalists to consider seriously that at this historical moment we live on a planet whose dynamics are "increasingly shaped by human activity—an anthropogenic Earth." Hence, humans must develop a more adequate ethical and rational capacity to manage the "coupled human-natural systems in a highly integrated fashion." [61] "The biosphere itself, at levels from the genetic to the landscape, is increasingly a human product."[62]

Allenby argues that the anti-technological thrust of much ecotheology is not only unrealistic but denies the human and religious dimensions of technology. Similarly, a technologist view that ignores the multidimensionality of ecosystems is extremely inadequate. The concept of ESEM (earth systems engineering and management) should guide a responsible ecological presence on the earth. The vision of a preagricultural, pretechnologic society that is inherent in much of environmentalism, including religious environmentalism, ignores the nature of humankind as a species. We evolved, Allenby reminds us, not only as a result of our genetic make-up but in conjunction with humanly constructed cultures. The earth also evolved as "a reification of human beliefs, especially those regarding technology, environmentalism, and religion,"[63] and ESEM is an attempt to take this into account. It is not a proposal for the construction of artifacts such as a toaster (Allenby's example), which presupposes an existing cultural and ethical context for which it is designed. Rather ESEM is more like designing the Everglades—in effect designing the context itself. The choice to create the Everglades is a choice to create a future in which humans and wading birds can co-exist. The ethical and religious dimensions "are not exogenous to the activity, but are important design objectives and constraints."[64] According to Allenby's view, based on the world we have created, humans are called not only to be authentic selves or even communities. Rather we now all live in and are morally responsible for an authentic world.[65]

Clearly, Allenby's view of human interaction with nature and the ESEM approach to ecological responsibility is quite different from those of the ecotheologians for whom cosmology and the organic interconnectedness of the earth provide the basis for envisioning ecological practice. There may be one vision that emerges as the most correct and the most effective, but our point for now is that the development of theological engagement with science is not at a standstill. It is a dynamic conversation that engages all who participate either directly or through the influence of the resulting texts. It challenges the prevailing values of the existing social imaginary of the West (at least) and provides a rich resource and ample room for a new set of ideas and practices to emerge.

6

GLOBAL AND LOCAL IN THE SOCIAL IMAGINARY

Isolating issues doesn't work because nobody lives like that.
—Vernice Miller-Travis[1]

Taylor's designation of a social imaginary as Western recognizes the particular contextual nature of the imagination in the construction of societies. First, this chapter examines some of the ways in which Christian texts have addressed the contextual nature of the ecological crises, focusing on ecofeminism and the evolution of its contextual richness. While air pollution, climate change, and toxic water do not respect national, city, town, or village boundaries, the ways in which these problems interact with local geographies, societies, cultures, and even race, class, and gender within particular geographies, societies, and cultures are recognizably different. Practices that we recognize as practices of hope are characterized by commitments that are exercised in particular locations. In other words, the social location of the participants matters and the participants themselves recognize their own concerns and visions of sustainable (and in our case, Christian) lives in the creation of the practice. Context enables the actualization of such practices.

Second, this chapter examines the theoretical discussions that reflect on efforts to relate the global and local dimensions both of the ecological crisis and of proposed solutions. There is no doubt that the social imaginary of the West, deeply shaped by Christianity, profoundly affected many societies around the world through colonization, in particular, but also through trade relations, international agreements, and many other forms of interaction both formal and informal. Today global economics and increased travel and migration are intensifying interactions among different ways of life and thereby intensifying the urgency of discussions of difference in all domains.

The significance of the relationship of global and local is heightened by the fact that other societies are influenced by the West and its social imaginary—particularly, for our purposes, Western Christian practices and theology. *Mujerista* theology, for example, is not just for Latinas, argues Ada María Isasi-Díaz. Rather it is a perspective on Christianity that ought to be taken into consideration by all liberation theology.² The argument has also been made for African and Asian women's theologies.³ With the rapid increase in global communication and travel, the influence of cultures and religions on each other can only increase. The postmodern awareness and emphasis on the difference between cultures, together with this increased interaction, make even more urgent the question of how best to confront the ecological crisis. Should we focus on global or local solutions? As Christian ecotheologians take up this question, they speak from a religious viewpoint that has a global reach whether or not their own social location is in the West. They are also aware of the concrete context within which Christians express their faith.

CONTEXTUALIZING ECOTHEOLOGY

The quotation from Vernice Miller-Travis that heads this chapter draws attention to the embeddedness of all life issues in the particularities of time and place. Miller-Travis speaks from an environmental justice perspective and draws examples from issues of environmental racism. Within that discussion she deals directly with the concrete local conditions within which people face ecological impairment and the difficulty of making their voices heard where national and international policies are formulated. This is as true for Christian and other religious groups as it is for the secular institutions. Miller-Travis reports having heard from colleagues working on global climate change in a large network of church groups in Kentucky. When she asked if members of the coalitions had spoken with coal miners and their communities in Appalachia she got the response, "We haven't gotten to them yet. We're dealing with the big picture issues. We haven't gotten down to that level yet. But when we get there, we'll invite them." Her reflection was, "How can you have a conversation at *any* level without the people whose lives are ultimately going to be destroyed by what we are recommending as fundamental policy changes, both domestically and internationally?" In Miller-Travis's view, "there is no big picture that is not essentially about how both nature and people can thrive interdependently *in each place*."⁴ She calls on the churches to support the process of bringing voices together. Her story exemplifies the notion of local concrete context.

Context does not always refer to such specific and concrete communities as the coal mining communities of Appalachia. It is a layered reality. In the first instance, ecological theology is contextual theology. The context (natural world, earth or the physical cosmos) is the content and central concern of ecotheology. Many call for more careful attention to the science of ecology, human behavior, and the evolution of the cosmos, and most ecotheologians do consider the importance of scientific findings and explanations. Then there are both theoretical constructions and practical lived understandings of nature that differ across cultures. The effects of ecological crises vary and are expressed in lesser or greater degrees across different cultures as well as in the societies, races, classes, and genders within those cultures. Rural-urban, town-village, agricultural-industrial, and so on, are factors that constitute the particularity of *each place* as Miller-Travis notes.[5] Christian ecotheological texts draw attention to one or more of these layers and sometimes to the concrete places themselves. One of the branches of ecotheology that exemplifies the gradual deepening of attention to contextuality is Christian ecofeminism for which gender, now in its many contextual settings, is the unifying concern.

ECOFEMINISM: EXPANDING THE CONTEXT OF GENDER AND ECOLOGY

Gender entered into the ecotheological discussion as more than awareness that all aspects of life are gendered. It entered an already politically charged form: feminism. Feminism is not only recognition of gender as a category that handily describes the different roles, duties, and responsibilities that people occupy and perform according to their gender. Feminism makes the claim that women are systemically oppressed, deprived of certain rights that accrue only to men, and are both blatantly and subtly excluded from many of the positions and activities that would give them power equal to men in society. Many feminists, as Anne Cameron has pointed out, did take stands and fight for ecological awareness and protection but were not labeled "ecofeminists."[6] Whether ecofeminism predated its labeling as such, ecofeminists see their contribution as unique. Janis Birkeland defines ecofeminism in a way that denotes its particular contribution to both feminism and ecology, as follows: "Ecofeminism is a value system, a social movement, and a practice, but it also offers a *political analysis* that explores the link between androcentrism and environmental destruction."[7] It is the connection between the oppression of women and the destruction of the earth that provides the common link among ecofeminists. It is also clear that from the beginning many ecofeminists were amply aware that change required transformation on many fronts—insights, analysis, political action, and practice.

The Beginnings of Christian Ecofeminism

With the emergence of ecotheology, Christian feminists, aware of the history of women within institutional church structures as well as the history of theological teaching about women,[8] were quick to see that the theoretic constructions of "nature" (meaning the physical world including human bodies) had a parallel with the issues and controversies surrounding women as full human beings. Rosemary Radford Ruether's essay, "New Woman and New Earth: Women, Ecology and Social Revolution," begins with the first argument made in ecotheology for the special consideration of ecofeminism in Christian ecotheology: "Since women in Western Culture have been traditionally identified with nature, and nature, in turn, has been seen as an object of domination by man (males), it would seem almost a truism that the mentality that regarded the natural environment as an object of domination drew upon imagery and attitudes based on male domination of women."[9]

Carolyn Merchant's work outlines the way in which this association of women with nature both influenced the emergence of Western science in Europe and was strengthened and handed on by science. Her work has been extremely influential for ecofeminists since.[10] An interesting version of the same motif is the use of the Adam and Eve story in the book of Genesis. Jane Caputi highlights the work of Michael Wood, film critic, in associating the gendering of nuclear bombs—their female names and the surrounding discussion of their powers of seduction and destruction—to traditional presentations of Eve.[11] Similarly, Elizabeth Dobson Grey highlights the use of gendered language associating females with situations of oppression or victims of violence, "virgin" forests to be "raped" for example.[12] Karen Warren gives a similar account based on journalistic reports of military language.[13]

While Ruether pointed out the ideological connection between oppression of nature and the oppression of women, she did not claim that women are *essentially* the same worldwide.[14] Her consistent contention has been that, while women share in gender oppression and may be seen in many cases to be closer to nature than men, this is due to the social construction of gender roles and not to some intrinsic affinity of women to nature. Many women, especially in the majority world, point out that there is a significant difference in the way women and men are present to the natural world. International agencies have taken this seriously and policies regarding development projects reflect this awareness.[15]

On the basis of her analysis of the role of domination, Ruether called for a transformation of society, including a "radical reshaping of basic socioeconomic relations and the underlying values of society."[16] Her work has been consistent in its association of ideological critique of how gender functions in society and a call for transformation based on concrete social action. Her progress in conversation with ecofeminist scholars, in nonreligious as well as multireligious settings, also mirrors and often itself initiates the introduction to Christian ecotheology of a growing awareness of the multiple representations of domination across class, cultures, race, ethnicity, religion, class, abilities, and so on.[17]

MORE ECOFEMINIST VOICES—DIFFERENT CONTEXTS

Rita Lester raises questions about the assumptions that underlie environmental discourse. In doing so she summarizes the increasing complexity of gender as a context for ecotheology: "The 'nature' environmentally concerned people work to 'save' depends on what understanding of nature is being invoked and for what purpose. What counts as nature or natural? Whose nature is protected and whose is not? When is nature a commodity? For whose children are we saving the world?"[18] Lester reflects on these questions in terms of four periods in the history of the United States, the colonial period, the nineteenth century, the Progressive Era of the early twentieth century, and the modern environmentalism of the latter part of the twentieth century. Each of these periods, she argues, presents a different set of assumptions regarding the questions she raises. These are assumptions that to a large extent still influence behavior today.

The early European colonists saw land as a gift from God to be controlled, in contrast to the Native Americans, who saw the natural world as a community to which humans belonged. Native Americans were seen by the colonists to be part of nature and not fully human. They were associated with wild animals. Likewise, African American women (slaves) were seen (and used) not only as labor but also as breeders of more slaves for their Euro-American owners. They were associated with domestic animals. Lester makes the case that these assumptions were deeply engrained, and racism continues to shape not only the progressive movements of the early twentieth century (e.g., eugenics) but also the environmental and feminist movements. History itself is a context. A type of feminism that sees itself as a universal category does not acknowledge the historical differences among women.

Just as feminism is not a universal category for understanding the ecological crisis and its effects, neither do racially constructed notions of nature apply across gender. Patricia-Anne Johnson and others have pointed out that while black theology did critique white racism, it virtually ignored the gender disparities within black communities and the disproportionate ways in which racism affects men and women.[19] Similar arguments can be (and have been) made with regard to ethnicity and class.[20] In a brief survey of a group of students entering an environmental studies class at the University of the Western Cape, South Africa, Ernst Conradie and Julie Martin found that while the students (most under 30) were aware of environmental concerns and had some general notions of gender, they did not see the connections of such concerns to gender, religion, and social issues.[21] It is reasonable to assume, as the authors of this study do, that there is still work to be done in articulating the relationships between ecological responsibility, other social concerns, and people's deeper values and convictions. This assumption is a prerequisite for ecological practice.[22]

The move from the white, middle-class context of feminism (understood as universal) to the recognition of different contexts for feminism came to widespread attention with bell hooks' *Ain't I a Woman?*[23] The title of her book, a quotation from Sojourner Truth, was a poignant reminder that African American women, and by implication nonwhite women everywhere, did not fit the assumptions of white European and North American women. In a sometimes biting critique, hooks uncovered the racism still perpetuated by the white feminist movement. "White women," she claimed, "saw black women as a direct threat to their social standing—for how could they be realized as virtuous, goddess-like creatures if they associated with black women who were seen by the white public as licentious and immoral?"[24] Furthermore, hooks pointed to the intricate relationship between colonization, slavery, sexism, and racism, historical realities that played out in contemporary American society.[25] Even the label "feminism" was seen as inappropriate, as connoting a certain mystique of the feminine that was not associated with most women of color. Thus womanism, the intellectual and activist solidarity movement of black women, was born and ecowomanism became the designation for that movement's interaction with the ecological movement.

As was the case with feminism, the womanist movement touched virtually all areas of the academy (and society at large) including religion and religious organizations. Further contextual refinements are represented by such designations within Christian studies as African Women's, Asian Women's, Latin American Women's, and *Mujerista* theologies. These theologies, far more than mere designations of the geographic location of voices, brought distinctive insights, concerns, and practices to ecotheology. Shamara Shantu Riley made this clear with respect to the emergence of ecowomanism: "While Afrocentric ecowomanism also articulates the links between male supremacy and environmental degradation, it lays far more stress on other distinctive features, such as race and class, that leave an impression markedly different from ecofeminists' theories."[26] In short, ecowomanist theology, as well as the other theologies mentioned above, breaks down the assumption that all women, or at least all feminists, experience the natural world in the same ways.

Other forms of ecofeminism and ecowomanism stress the cultural ethnic context, and bring into conversation with Christianity the ideas, rituals, and practices of other traditions. One of the most graphic presentations of this kind of both conversation and also confrontation of different traditions occurred at the 1991 meeting of the World Council of Churches in Canberra. Chung Hyun Kyung, a Korean feminist theologian, gave a keynote address. In it she evoked in a traditional Korean ritual the spirits of Korean ancestors and equated the spirits with the Holy Spirit of Christianity. Some Council participants did not appreciate the equating of Christian beliefs with "paganism." In her later publication, *Struggle to be the Sun Again*, Chung provides an illustration of the complex interaction of Christianity and other religious traditions in a Korean context. She relates the story of a Korean mother whose son was killed by a powerful politician. The court subsequently declared the killer innocent even though it was generally known that he was guilty. After distributing leaflets about his guilt, the grandmother of the slain boy was imprisoned for libel. Out of her desperation for justice, the slain boy's mother drew upon all the resources she had. She drew portraits of the politician and his powerful supporters. Each day she prayed in front of a bowl of pure water to all the deities she knew. Then she pierced each of the portraits with an arrow. As news of her actions and prayers spread throughout the town, the politician and his cohorts became extremely fearful; in the shamanistic tradition such actions would lead to their deaths. The grandmother was released from prison and the family compensated for the unlawful death of the boy.[27] As Chung concludes, "In this powerless mother's story is seen a glimpse of Asian women's ecumenical spirituality. . . . She asked for help from the supreme Korean god, a god from the shamanistic tradition, and from Jesus."[28]

Similarly, in terms of ecotheology, Chung draws attention to the specific ecological meanings that various Asian traditions add to Christian understandings. One of these is the traditional Filipino worship of the great Mother, *Ina*, the Divine Womb from which come forth all forms of life. This image has become conflated with the image of Mary, Mother of Jesus, who is worshipped extensively throughout the Philippines.[29] These are examples of ways in which non-Western contexts colonize a Western Christian social imaginary.

Latin American ecofeminist theology owes much to the feminist theologians who initiated and/or became members of *Con-spirando* (breathing together), a liberationist and ecofeminist collective in Chile; the collective publishes a magazine by the same name.[30] This collective and its publications are illustrative of the interweaving of theological meaning with cultural practices, which have characterized much of ecofeminism from the beginning. Judith Ress articulates the central argument of this book: change is most likely to happen in the gradual building up and spreading out of ecologically responsible cultural practices informed by ecologically poignant meanings. She writes of *Conspirando*: "Over the years, we have come to see that our ability to influence the status quo is in advocating for cultural over political change. . . . We also find ourselves less attracted to massive mobilizations and protests, preferring small groups and local issues as our locus for weaving new visions and nurturing energies to edge these visions forward."[31]

The collective and its members are committed to ecological sustainability, to "local initiatives that enhance the bioregion" including composting, recycling, eating organically locally and seasonally as far as possible, and other similar practices. They also actively support all efforts to promote better environmental policies and cleaner more sustainable development in Chile as a whole. Ress notes that there are now a growing number of grassroots women's groups in Latin America with similar commitments and an ecofeminist vision, most of them emerging from women of faith. While some groups are still small, there is a growing dynamic network.[32]

Reports on the work of *Con-spirando* also clearly highlight how contexts of ecofeminist Christian theology overlap and influence each other across many borders. Not only are class, race, ethnicity, religion, and gender borders crossed within particular geographic contexts,[33] but international and regional borders are also transgressed. Judith Ress honors the contribution of Ivone Gebara, a Brazilian ecofeminist theologian, in bringing the ecofeminist vision to *Con-spirando* and to Latin American theology in general.[34] Gebara in turn credits Rosemary Radford Ruether, a North American theologian, as introducing her to Christian ecofeminism.[35] Ress gives a more complex account of the indebtedness of *Con-spirando* and Latin American ecofeminist theology to Latin American and international theologians, ecological movements and philosophies, in general, in her account of ecofeminism in Latin America. She interviewed 12 Latin American feminist theologians about their personal spiritual journeys.[36] All describe their journeys in terms of some substantial interweaving of the contexts of political and social realities in their regions and countries, particular aspects of Christian (mostly Roman Catholic) heritage, such as honoring of Mary, that are characteristic of Latin American spirituality, Latin American traditional cosmologies (e.g., Aymara and Quechua), and international ecological movements. The work of Christian ecofeminists is characterized by the way particular contexts shape their visions and actions.[37]

A quite different context for gender and ecology is offered by Dorothy Soelle. Extremely conscious in all her writings of the inheritance of the Holocaust and its implications for post-Holocaust Christianity, Soelle points to the necessity of understanding and incorporating liberation theology and its focus on redemption into the theology of creation.[38] It is not the natural world, she argues, that releases all kinds of slaves. It is also the case that imperialistic constructions of society have based their ideologies on notions of the cosmic order. Liberation, including women's liberation and the liberation of creation from human domination is founded, in Christian terms, in the redemptive power of the life and death of Jesus Christ. It is this focus, Soelle claims, that should ground a new Christian understanding of the God-human-earth relationship. The emphases on God as Lord and on the spirit-body dualism of humankind have distorted the human-earth relationship and relationships among humans. Soelle calls us to a realization of the creative tension of human existence. "We are created in God's image, but we are made from dust."[39] Both of these must be held in lively dialectic. Likewise, creation ought not to be viewed as an act of the past only, but as an ongoing reality—creation is also the future. What we do now in large part contributes to the manner of ongoing creation.[40] Set in the context of memory of the Holocaust as a stark human failure, Soelle's is a persistent reminder of human responsibility for history and the fact that the present ecological crisis is a historical reality to be confronted by human action.

RELATING THE GLOBAL AND LOCAL

The particularity of responsible ecological practices is not to be equated with eccentric isolationism, nor with the nonpolitical. For ecotheologians, engaging the particular is the methodology by which the whole is transformed. As coordinator of the World Council of Churches team on Justice, Peace, and the Integrity of Creation, Aruna Gnanadason illustrates the necessity of paying attention to the particular. She bases her argument on the ecological history of India related by the social scientists Gadgil and Guha. In examining some of the ecological effects of British colonization of India, Gadgil and Guha discovered that the Christianization of India during colonization resulted in the eradication of many traditional practices beneficial in the care of natural resources. Remnants of these practices still exist especially among indigenous populations in parts of India. These include the existence of protected Sacred Groves and quotas on forest extraction. "If this [destruction of traditional practices] is our heritage as Christians in India," Gnanadason reflects, "then the imperative to redeem our history is urgent."[41] She claims further that this redemption can take place only if we pay attention to the particular, however strange and new. In her words:

> We now hear new voices of hope from women, from the Indigenous peoples of the world, from groups considered inferior such as the Dalits, and from the earth itself. These are strangely new voices, but they have to be heeded. There is a call for a major paradigm shift, a journey into the unknown and the unfamiliar. This is a journey into the, until now, feared local and particular. This is a journey of being attentive to the place and the memories that go with it and the experiences that give it and the community, identity.[42]

In her own reflections, Gnanadason relies as well on the works of other ecotheologians such as Gabriele Dietrich, Sallie McFague, and Larry Rasmussen, all of whom call upon and reinterpret relevant teachings of the Christian tradition to give religious meaning to the often "feared local and particular." The basic incentive she finds articulated by Douglas John Hall as "To turn towards God (*theo*centrism) was simultaneously to turn towards God's beloved world."[43] Following Sallie McFague, Gnanadason reinterprets the Christian God as a Lover, passionately caring for the whole Earth, and as Mother, the feminine image of divine activity of creation and justice. To these, she adds the very particular (and strange to Western Christians) "God as the resisting Dalit woman—[image] of God as one who resists injustice and who protects the earth."[44] These particular actions in very concrete contexts are hopeful, not only for the small places in which they are enacted, but also for universal systemic change.

The embodying of ecotheology in concrete contexts is not without controversy and difficulty. As Maria Pilar Aquino points out, in Latin America women's religiosity is often quite distinctive and constitutes a significant expression of their Christianity. While it is part of their cultural identity and very important to consider within their context, it contains troubling elements that legitimate aspects of their own oppression, as well as memories of resistance.[45] Many women seek the comfort of a spirituality that enables them to accept and conform to unjust strictures of family and community. A second issue arises from the global nature of many ecological problems. Air and water pollution, climate change, and extinction extend beyond local borders and are often difficult to even notice in local contexts. A third issue, and perhaps the most difficult, is the rapid pace of economic globalization. Ecotheologians are beginning to respond to the growing awareness of the ills accompanying economic globalization. They question whether their own efforts to date have been helpful in curbing the rapid social and ecological devastation that accompanies economic globalization.[46] Does the systemic power of multinational corporations just overwhelm any local efforts to contain it? In raising the issues involved in the global/local relationship, we are asking in yet another way about the effectiveness of ecological practices.

When ecotheologians emphasize the importance of contextualization and increasingly contexualize their own insights, most are not taking sides in a dualism that pits local against global in relation to any of the three dimensions of the global/local discussion. The weight of present ecotheological texts lies with the view that local action, as transformed practice or actions of resistance, has the best chance of making a difference in a crisis with global dimensions. Certainly this does not mean just any and all local action. Irrigating a field may benefit a local village temporarily but may be quite damaging to a larger water system.

Attention to place is a key criterion for the engagement of local practices with ecological concerns. This in turn informs and strengthens the arguments of ecotheologians about attention to particular contexts, as outlined above in the case of ecofeminists. It also provides a way of addressing the relationship of global to local issues. In drawing attention to the importance of place in transforming how we think about and relate to the natural world, Kimberley Whitney investigates the multiple meanings of place to include the following: "*Place* is a heuristic tool that begins with, or moves from, embodiment, where one wheels, sits, stands, walks in space and time. In this conceptual frame, persons and communities are 'implaced' or located in particular socio-historical and physical (ecological and geographic) locations. Place is a fragrance or touch or gaze that includes both physical and social factors."[47] Among the physical and geographic aspects of place are land, animals, air, water, and human architecture and infrastructures. Social and cultural dimensions include "politics, economics, histories including hidden or subjugated histories, worldviews, moral and religious imagination, the arts."[48] All of these aspects give rise to and "spur the practices and spatial patterning of place and world-making." They are, according to Whitney, "*constitutive particulars*" of place.[49]

But Whitney is not interested only in the description and definition of place. Her real concern is to address the link between the global and local in matters of ecology and justice. How, for instance, does one link *place* to the ethics of sustainability or the particulars of the local with the concerns of the global? Global imbalances of power are often mimicked in the local context, and local communities often have their own imbalances. Global ecological precepts and policies are often grounded in one particular, more powerful reality that is in fact local but is assumed to be universal.[50] This is a persistent commentary on the Western bias of most so-called international strategies.[51] In addressing such critiques of the local, Whitney points to the permeability of the boundaries and the nestedness of the local. "I must interrupt," she says, "a mired conversation [regarding the relationship of global and local], arguing it is vital to make strategic distinctions between 'local' and 'global,' 'particulars' and 'universals' while at the same time recognizing the nesting into which local particularities *unfold from* and *furl into.*"[52] In lived practice, local and global, particular and universal, can and most often do impinge upon and enter into the construction of each other. Place gives a concrete ground to this relationality.

In similar fashion to Whitney's theorizing about the significance of place, Adrian Ivakhiv focuses on the relationship of universals to concrete performances consistent with those universals. He writes:

> Practice is, of course, never entirely separable from theory, that is from idea, belief, discourse, ideology, and cosmology. But while scholars of religion have all too often focused their attention on beliefs and values, the crucial point here is that these are always expressed and given shape in and through practices, forms of engagement and relationality in which socially constituted groups take up the world around them-

selves in different ways, shaping and re-shaping it in the process. The emerging focus on performance, embodiment, "lived religion," and "religion in practice" . . . shows that anthropology's growing concern with these topics has not eluded religion.[53]

Again, it is only in the concrete context of place with its contingent social and cultural meanings, its particular construct of class, gender, race, ethnicity, and so on that the global and the local interact and progress is made or not. As mentioned above, local practice is not necessarily good, that is, ecologically effective as well as attentive to both human and beyond human suffering, in this case. In much of the writing of ecologists, and certainly in the case of the Christian texts under consideration here, the question of ecologically good local practice is a primary concern. Contextualization, paying attention to the particular, the local, the ways the ecological crisis itself, as well as the responses to it, are embodied—an essential part of what they consider effective ecological practice.

Both Whitney and Ivakhiv tie the local to concrete places where people live with their mixture of values and identities making a world together in the ordinary (but certainly not insignificant) everyday-ness of life. This is in contrast to what one might consider the understanding of local groups that assemble to focus intentionally on a particular aspect of life—environmental groups such as the Sierra Club, the World Wildlife Fund, or even a student activist group. However, these groups have significant value and may well constitute important commitments within the kind of local contexts of which Whitney and Ivakhiv are speaking.

In *Turning the World Right-Side Up*, Patrick Kerans and John Kearney argue for the critical role of neighborhoods of which Whitney and Ivakhiv speak. Neighborhoods are key to the building of community, which in turn is necessary for resistance to globalization. There are three major reasons that neighborhoods are key to this resistance: (1) unlike groups that come together for a particular cause, members of neighborhoods usually represent many sides of an issue and accommodations must be made; it is not easy to walk away. Such listening and accommodation demand and build skills in real dialogue. (2) People in neighborhoods live in distinctive ecosystems, which in one way or other enter into the decisions. (3) Most importantly, coming together as neighbors is in the context of day-to-day life, as opposed to coming together as members of some club or organization. The actual coming together is what makes the future livable in that particular neighborhood, and to this end solutions must be integrative of many factors and must be sustainable, both ecologically and socially, if life is to continue in that local place.[54] For Kerans and Kearney, this is necessary not only for sustaining life in local places but for the creation and maintenance of viable and sustainable structures of governance at both the national and international levels.[55]

One finds this same argument applied to different cases in many religious texts as the ecotheologians attempt to bring their disciplines, beliefs, values, and activism to bear on the resistance to globalization. While the term globalization has many uses and its meaning is often ambiguous, we use the term in an economic sense, referring to the reality of the worldwide reach of multinational corporations enabled by free trade agreements and advancements in communication technology, among other factors. There are, no doubt, arguments to be made for some benefits of globalization, but the focus here is on efforts to confront ills that are at least exacerbated if not caused by the contemporary global spread of corporate capitalism.[56]

Gnanadason states the issue in this way: "We [World Council of Churches] need to situate our efforts to care for the earth in the context of the work being done now of challenging the project of globalization and the trail of unsustainability it leaves behind."[57] Heather Eaton and Lois Lorentzen introduce their anthology on ecofeminism with this question: "How does ecofeminism confront the many faces of globalization?" Among the faces they include deforestation, wars, militarization, and socioeconomic impoverishment.[58] Eaton is representative of many ecofeminists who make impassioned pleas to consider how ecofeminists (in particular Christian ecofeminists) can make a difference for those who suffer most the effects of globalization. In her words:

> The players of imperialist globalization or the World Bank do not care whether Wisdom/Sophia was present in Genesis, or that the Christian eschatological doctrines are distorted. But they will care if Christians, inspired and empowered by a new understanding of Wisdom and eschatology, resist the efforts of the World Bank, the World Trade Organization and the International Monetary Fund to prevent ecological ruin. . . . Thus, to evaluate ecofeminist theological work, one measure is whether the approach can move the paths in effort into the work of resistance, vision and reconstruction.[59]

This is the challenge of all ecotheology.

In the conclusion of his work, *The Environment and Christian Ethics*, Michael S. Northcott highlights instances of effective resistance and action in local communities. Fishermen on the Hudson River lobbied power utilities and relevant governments over the pollution of the river and took their cause to the local courts, with positive results. Local people's protest in East End London resulted in the closing of a highly polluting power station in their neighborhoods.[60] Northcott is also clear in his convictions that the Christian churches ought to encourage awareness of the global realities associated with the ecological crisis, as different localities are affected differently with special concern for the poorest places in the world. However, he is critical of the discourse associated with global images of the Blue Planet. Large industrial corporations often use this image to advertise their earth friendliness; the Bruntland Commission and other international agreements tend to enforcement of global solutions on local commons.[61] Arguing against the view that place-based communities will quite likely act in their own interests and not in that of the whole earth, Northcott points out that "it is modern nation states and their colonial and corporate frontiers that are most commonly responsible for encroachments on the land rights and common property procedures of traditional place-based communities, not other place-based communities."[62] Northcott calls on Christian churches to invoke their parochial traditions in order to be effective actors and agents of empowerment within communities of place rather than in gatherings of like-minded persons. He claims that communities of place embody the realities of face-to-face community life that make dialogue and action an effective counterpoint to any utopian global management of the earth.[63]

Many Christian ecotheologians think that concrete local and contextualized practices are the most effective way to confront the ills of globalization, including the ecological crisis. The network of local efforts is the basis for hope. Larry Rasmussen proposes a framework that addresses both the social and ecological issues that arise from the contemporary rush to globalization. The framework involves transitions at all levels and through all dimensions of life: economic, social, institutional, informational, demographic (population), technological, moral, and religious. While these transitions call for international, cross-cultural, and inter-religious collaboration, they are not the vision of "sustainable development," the rhetoric often used to green the growing global economy. In Rasmussen's graphic description, "Such a course [sustainable development] assumes, and leaves, the powerful forces of centuries-deep globalization in command. The result will likely be a hard, brown world tinged with soft, green edges for the rich and aspiring rich. Trying to wrap the besieged environment around the ever-expanding global economy is not the way to go.... Rather the way to go is sustainable communities."⁶⁴ Likewise, having described the efforts of communities in West Harlem to fight the growing pollution in their neighborhoods, Peggy M. Shepard challenges environmentalists and technocrats who are attempting to deal with ecological problems to support the "growing sense of ownership" of environmental and justice movements at the grassroots level.⁶⁵

As we applaud the efforts and actions of neighborhoods around the world, it is well to return to the reflections of the ecofeminists with which this chapter began. Contextualization means that the context is really part of the action. It is one thing for middle class (not to mention wealthy) Christians to advocate simplicity of lifestyle and to make it count as an operative value in the negotiation of their neighborhood setting. This is not to be considered insignificant and may be critical to all peoples around the globe if the consumerism of the richer nations is to be curbed. It is quite another thing, however, for subsistence farmers in any part of the world to consider the reduction of pesticides on their crops. As Anne Primavesi observes of those of us in the richer nations, "We no longer work in order to eat in order to live."⁶⁶

Christian and ecological quests for hope ought never romanticize the lot of those in the world who live the most lightly on the planet. By keeping together the various contextual components that characterize both the devastating effects and the promising solutions to the ecological crisis, ecotheological texts are a necessary irritant to any attempt to romanticize parts of the world where "the goal is not resistance, but simply existence."⁶⁷ Cynthia Moe-Lobeda reflects on her own position as one of economic privilege but where she also hears the call to radical change of her available lifestyle: "From this location, I am also dislocated—displaced by having stepped into the lives of people brutalized by the economic structures that provide my consumer privilege. . . . Commitment in community with others to resist the culture of consumerism, excess, and economic exploitation furthers the dislocation."⁶⁸ The contextualization of ecotheology brings to practice not only the concreteness of place within which all practice takes place, but also the critical awareness of the nestedness of contexts and local places within other contexts and places. Practice that is just and truly ecological is suspicious of the dynamics of power that lead to the universalizing of one place or context over many or even all others. For those whose context and home include Christian faith, the ecotheological texts help in the reimagining of life together in very localized settings, which are also global in their reach and influence.

7

LIVING *AS IF*

This chapter considers *lived* practices of hope. Written texts can ultimately be judged to be effective practices of hope when they demonstrate that they hold the seeds of change that influence a social imaginary. Ecotheological lived practices that continue to reshape an ecologically conscious social imaginary are, in this sense, effective practices of hope, influenced by the texts we have been considering. In this chapter we look at a sample of inspiring renewed practices in a variety of congregational and denominational settings that have energy and promise to reshape a social imaginary toward an embrace of environmentally sound awareness and action.

In *Gaia's Gift* Anne Primavesi makes an important link between personal revolution of self-perception and awareness of our evolutionary history. She writes about "living *as if* we understand and give priority to the complex range of interdependent relationships on which all life here depends and in which we are totally involved." This is a revolution "in which one lives in a far from ideal situation 'as if' in an ideal one." This can come about by "looking back though our species' history and seeing ourselves as belonging to a longer and older ancestry [living] *as if* we are what we have always been—members of the community of life on earth."[1]

This chapter uses evolutionary psychology, evolutionary biology, and congregational studies to preface and frame our study of intentional groups as living texts of hope. From evolutionary psychology we argue that the default position of groups is to take care of their own. This impulse to take care of one's own may seem to work against the impulse to care for the larger world. We would instead like to declare it a point of departure for change. From evolutionary biology we consider the Grandmother Hypothesis, which helps us make sense of the way in which aging congregations and intentional gatherings of elders can function to preserve the species. Both social science literature on grandparenting and theological literature on aging as a countercultural vocation give contemporary insights into the role of elders in environmentally sustainable practices of hope. We also use insights from congregational studies. We explore resources that identify the eras in which churches were built.[2] This is important because, as is obvious from observation, it is easier to build green than it is to rebuild or renovate green. We examine the argument that congregations are the appropriate places for environmental action because of their structure, size, and ways of operating. Lastly, we look at recent literature on congregations that argues that members must be part of two or three groups: a group of 15 or fewer, a group of 50 or fewer, and, in the case of large churches, one of 150 or fewer. These studies link closely with findings from evolutionary psychology. Together, evolutionary and congregational insights shape the context within which we present living texts that are practices of hope.

EVOLUTIONARY PSYCHOLOGY AND THE PRACTICE OF HOPE

As indicated in chapter 5, Carolyn King argues that it is easier to understand human nature if we understand biology; if we understand that "culture is an elaboration of biology"; and if we recognize that "free will must be exercised within biological constraints."[3] This understanding of human nature requires a reexamination of how churches function to influence the practices of their members. From the perspective of what evolutionary psychology has to say about the evolution of public morality and about how community decisions are made,[4] we learn specifically that "the default setting of human nature . . . is the need and ingrained habit of fitting in with, and looking out for one's own closest group."[5] Few biological predispositions are as real as the need to care for one's own group and most of all for one's own children—and we judge this to be right. "The force behind the moral attitudes that comes most naturally is to be conditionally co-operative with member of one's own group, but much less co-operative with—at least wary of, and necessarily hostile to—members of other groups."[6] But this is not enough for earth in crisis. King concludes, "Unfortunately for the environmental movement, it is basically unnatural to humans to think in terms of the global, rather than the local, community."[7]

Ecotheologians and eco-ethicists are trying to evoke a moral response to environmental crises. What does it mean for the possibility of a moral response that taking care of one's own group has the most natural force behind it? What can mitigate the power of this default position? At the level of the individual, a person cannot escape at least some of the negative results of anti-environmental behavior—individual contribution to smog pollutes the air we all breathe. But this is not motive enough to change behavior. As game theory demonstrates, in a situation of diminished or scarce resources, the default moral position is to get what one can before it's all gone. Or to use another analysis, random positive reinforcement trumps negative reinforcement. What about cultural selection as it influences a group? While natural selection favors individual attitudes that have been adaptive in the distant past, "cultural selection operates as an adaptive accumulator of memes (ideas), affecting groups as well as individuals and involving conscious motives and moral agency over the short term."[8] Humans have a limbic system that controls emotions and "stimulates the conscious feelings of love, fear, racial hatred, sexual jealousy, and many more that profoundly influence our daily behavior."[9] Yet these feelings do not overcome deep-seated natural emotions that keep us focused on our own small group. "All emotions that help to bind a group together and distinguish it from other groups are strongly favored by cultural selection."[10]

If not natural morality, what? Evolutionary psychology allows us to detect and reject ancient prejudices. Seeing them, we can resist being manipulated by them. For a Christian who takes theology seriously and who understands it from the heart, realities called sin, grace, and sacraments are powerful resources.[11] "Cultural traditions such as religion reinforce existing group selection favoring the better survival of the most strongly cooperative social groups. It is our best hope for the future."[12]

> In so far as religion can encourage group-centered, cultural-based altruism at the expense of personal selfishness, which in turn favors the genetic survival of the group, then religious traditions can be a cultural force which favorably influences the direction and/or strength of biological evolution.... Such an attitude requires the future, thinking church to accept as valid a rather deliberate sort of faith.[13]

Such a deliberate faith does not deny the human character of religion, nor does it think it any less vital for that. "Biological predispositions are real, but they cease at the frontier of the tribe, so that is where Christian theology must take over from cultural evolution [to answer] the greatest question of all: 'And who is my neighbor?'"[14]

As is obvious to even the most casual observer, many if not most churches in the United States and Canada[15] have populations that are older than the age distribution of general demographics would indicate. King's analysis adds a positive perspective to this.

Older adults have typically been together for a long, long time. They have aged in place. Younger families and their children may have come and gone, but the old-timers are still there, still talking to each other, still supporting each other in groups with poignant names like "The Young Adults Class." By the longevity of their lives together, these groups meet conditions that King sees as necessary although not sufficient for new and sustainable thinking about the environment: "Only when self-interest is restrained by local interaction and the relentless scrutiny of inescapable close associates can it drive the various forms of cooperation and conditional altruism that underpin the lives of social animals."[16]

As social animals we are adapted to small groups (150 or fewer). A danger of the ecological crisis is that we will retreat to protectionist strategies to preserve our own groups. The churches can play a role in mediating a more appropriate, effective, and moral solution. They can do this only if they gain and maintain the respect of all parties, meaning that "modern churches will have to recover something of the Christian claim to have something rational to contribute to the environmental crisis."[17]

Current congregational studies note, without any reference to the literature of evolutionary psychology, that the building blocks of congregations are the "Small Group/Sympathy Group" of 15 or fewer, the "Primary Group/Family Group" of 50 or fewer, and the "Community Group/Village Group/Fellowship Group" of 150 or fewer.[18] One may observe in congregations where older adults predominate that congregants do quite regularly belong to a group of 15 or fewer (a Circle, a Sunday school class, a Men's group, etc.). Many older adults worship in the half of all congregations in the United States where there are fewer than 100 members in worship. These groups are often long-lived and do potentially provide the dynamic of reciprocity of supportive social groups to which King refers. Of course, these groups can also function negatively, precisely because of their cohesion, to exclude new members from joining.

Twenty-five years ago those who are now elders in our churches "enthusiastically supported the policies that still dominate the planet."[19] According to John Cobb Jr. there is now widespread disillusionment and openness for new ideas. There are a variety of reasons for this, including the dynamics of the phenomenon of age itself as a factor in a wider vision.

> In old age something special happens to reality. Its hardness is softened by the experience of transitoriness. . . . Reality then becomes questionable—not as in youth, when time seems endless, but rather because now reality has been found not to be as real as it appeared in the realistic period of mature life. The view of things widens out. Under pressure of reality, a person was limited to the present moment. But toward the end the whole comes again into view. As in autumn, when the leaves fall from the trees, the view expands, and one is conscious of wide space. Reality engages the will in what is at the moment to be sought, done, mastered. But as the years go on one learns to loosen one's hold. The urgency of the will begins to slacken. Detachment is the next phase, and a person's nature opens up to the whole, to a general view of existence.[20]

For Cobb, this is a hopeful condition for the embracing of an ecological ethic and practice.

One dynamic of the environmental crisis is the challenge to our comfortable lifestyles. As King notes, "Few people are willing to face the unpleasant fact that, sometime quite soon, it will no longer be possible to carry on with our lives on the assumption that the future will be a more or less logical extrapolation of the present."[21] Inevitably, age brings with it awareness that one is closer, and then much closer, to the end of life than to the beginning. One's future cannot possibly be a more or less logical extrapolation of the present. In the face of an awareness of the transitoriness of existence, detachment can provide openness to concerns, including concern for the earth community generations ahead. It is our contention, of course, that this shift to a wider vision caused by aging has at least in some instances been augmented and intensified by the sustained efforts of the ecotheologians and ecoethicists who have for all these years been teaching and practicing hope. Their texts can make plausible the practice of living passionately with diminished resources and a diminished future.

EVOLUTIONARY BIOLOGY: THE GRANDMOTHER HYPOTHESIS

There is a recent part of our evolutionary history—just 30,000 years ago—that we still benefit from today, namely the care of grandmothers, who have raised children while their own adult children were hunters and gatherers, fisher folk and farmers, artisans and miners. Compelling in its urgency to create a safe future is the contribution of grandmothers, who in their connection with their grandchildren display a deep, somatic, and even survival-of-the-species kind of faith. This must be understood within the larger context of history and the survival of the human race. There must be a reason for menopause.[22]

This is not simply an accident of evolutionary history. "Twenty or so vigorous years between the end of reproduction and the onset of significant senescence does require an explanation."[23] From an evolutionary point of view, longevity is not an accident but is somehow crucial to our development as a species.[24] That's the perspective of the so-called Grandmother Hypothesis, which suggests that the post-menopausal survival of grandmothers permitted the nurture of young children and the success of the human species by providing extra care for children. The first grandmothers, in the Neanderthal era, were available to babysit, and while they did so, they had the time and the audience to which to pass on stories from their own youth. This is credited with sparking the birth of traditional culture. Grandmothers are believed to be the evolutionary advantage of modern humankind.[25] From a sociobiological point of view, there is "accumulating evidence to suggest that post-reproductive women can indeed have a significant and positive effect on their offspring's reproductive success, . . . and hence forward more of their own genes to the following generation."[26] Current research points to a biological rather than cultural explanation for the Grandmother Hypothesis. The authors of "Grandma Plays Favourites: X-Chromosome Relatedness and Sex-Specific Childhood Mortality" reevaluate the grandmother effect in seven previously studied human populations [. . . and] demonstrate a relationship between X-chromosome inheritance and grandchild mortality in the presence of a grandmother. Their evidence indicates that grandmothers favor the grandchild that carries their X-chromosome. Built deep into the unconscious mechanisms of the brain is a level of attraction and bonding that helps grandmothers do what they have done (and in most parts of the world continue to do)—care for the grandchildren while the parents toil in the workplace.

If our experience, because of social class, is of leisure or companionship grandparenting,[27] we may see only the emotional and sentimental links between grandparents and their grandchildren. However, we have access to a historically more typical picture of what grandmothers have always done if we look outside middle- and upper-middle class social circles. Even in our own developed countries, many grandmothers raise grandchildren because it is their cultural norm, or because parents are absent by reason of work, sickness, or incarceration.

The evidence for the role of grandfathers differs, at least in the research of Virpi Lummaa, a historical biologist who studied Finnish church records from two centuries ago. Her research and that of her colleagues showed that grandmothers "provide direct aid in ensuring the survival and reproduction of their grandchildren. The same records showed, however, no such benefit from fathers and grandfathers." She goes on to say that a living grandfather did not increase the number of grandchildren. "If anything there's a negative effect," she concludes. This could be because of the cultural tradition of catering to men, particularly old men. "Maybe if you had an old grandpa, he was eating your food," she speculates.[28]

What is the role of grandmothers and grandfathers in the United States? More specifically, how many grandparents live with and/or are responsible for their grandchildren? There are two major data sources: the 2000 U.S. Census and the U.S. Census 2005 American Community Survey. In October 2003 the U.S. Census Bureau issued a Census 2000 Brief, "Grandparents Living with Grandchildren," written by Tavia Simmons and Jane Lawler Dye.[29] This factual and narrative report identifies 5.8 million households (3.6 percent of 158.9 million) where grandparents were coresident with grandchildren under 18. "Among these coresident grandparents, 2.4 million (42 percent) were also 'grandparent caregivers,' defined in this report as people who had primary responsibility for their coresident grandchildren younger than 18. Among grandparent caregivers, 39 percent had cared for their grandchildren for 5 or more years."[30] Of all coresident grandparents, 40 percent were women.[31] "The proportion of grandparent caregivers living in poverty [was on average 19 percent]. The proportion of grandparent caregivers living in poverty was highest in the South (21 percent), and in some states it was as high as 30 percent."[32] Living in poverty is a specific term, meaning to live below a government-designated poverty line. As James Sykes points out, however, "In the U.S. there are 2.4 million households headed by grandparents rearing grandchildren. Of those, 78 percent experience financial hardship and 52 percent do not have enough resources to meets their grandchildren's needs."[33] Lastly, as the American Community survey notes, there is wide variation state-to-state of the percent of coresident grandparents who are caretakers: state totals range from a low of 24.2–34.1 percent to a high of 58.8–63.9 percent.[34]

Grandparents raise grandchildren because, we presume, they have to.[35] "Have to" does not suggest only practical necessity. At least if the Grandmother Hypothesis has any validity, grandmothers have a much deeper stake in the lives of their grandchildren. At the simplest level, the healthy lives of the grandchildren means that the grandmother succeeds in passing on her genes. But at a more complex, and perhaps harder to prove although often observed level, grandmothers raising grandchildren have a deep faith in the future of these children; this is the root and raw energy that wills an older generation to give all the young a chance for a sustainable future. This is a practice of hope in often quite desperate circumstances, as "so many elders, especially grandmothers, face daunting tasks every day and live with constant pressure and uncertain futures."[36] Grandmothers such as these who use "their role to change their families and society for the better [provide a template that] is being extended to all elders for the establishment and implementation of a new and powerful social identity."[37] Given the aging of Christian congregations, given the evolutionary role of grandmothers, given the long-lived character of relationships in aging congregations, and given the phenomenon of what age itself brings in the view of existence, the possible role of grandmothers in green hope should not be underestimated.

Evidence that Baby Boomers (the generation born 1946–1964) give more to charity than pre-Boomers (born before 1946)[38] does not invalidate our suggested conclusions for two reasons: first, the difference is quite small: $1,361 to $1,138; and second, the largest group of caretaker grandparents are people aged 50–59.[39] Somewhat disquieting is a report of a "survey of 14,000 respondents to a Eurobarometer survey." There is a steady decline [in support for climate change] with age, based on whether respondents are willing to pay a higher gasoline tax.[40] This phenomenon, as reported, needs careful tracking in other countries.

CONGREGATIONAL STUDIES

As we shall demonstrate, there are Christian (as well as Muslim, Jewish, and other) congregations intentional in "mobilizing a national religious response to global warming while promoting renewable energy, energy efficiency and conservation."[41] A study of Christian congregations from various perspectives helps construct a picture of green hope that is made actual, physical, economic, and often political. At the core, one discovers a growing sense of congregational covenant to make a difference in the current ecocrisis.

Aging Physical Structures: Challenges

Many of these congregations face the burden of buildings constructed long before consideration was given to energy efficiency. Indeed, the median date that worship began at present location for all congregations in the National Congregations Study (USA) is 1958.[41] One pastor interviewed about the sanctuary in which her congregation worships referred to the church facility wryly as "our albatross." Nor is the burden of old buildings evenly distributed among denominations. For example, a comparison of mainline and evangelical churches reveals that while 70 percent of evangelical congregations in the United States were founded after 1975, only 10 percent of mainline Protestant denominations were founded that recently.[42] Old physical plants may be less of a burden to other institutions—for example, in a growing school district a new elementary school can be built green for teaching as well as energy saving purposes. On a larger scale, Wal-Mart can effectively demand of its suppliers less wasteful packaging. These are not the kinds of changes available to most congregations.

There are, however, singular instances of renovation of large structures. In Monroe, Michigan, a Roman Catholic religious order for women renovated their buildings with environmental commitments in mind. The average age of the sisters living in this Motherhouse is 86.

> The decision to reduce our dependence on non-renewable energy sources resulted in a geothermal system for heating and cooling the 376,000 square feet of the Motherhouse. . . . In spite of adding 305 new bathrooms to the Motherhouse, using high-velocity, low-flow fixtures and fittings and a graywater flushing system, we have reduced overall fresh water consumption by 49.6 percent. In addition, the use of vegetated swales and wet meadows enables us to divert almost a million gallons of water per year from the municipal storm sewer system. A separate piping system collects used water from sinks and showers in the Motherhouse. This is called "graywater." The pipes route the water to a constructed wetland on the campus. The constructed wetlands, mimicking nature's purification system, cleanse the graywater and recycle it back into the Motherhouse for flushing toilets.[43]

Another remarkable instance of renovation of an energy-inefficient building is St. Stephen's Cathedral in Harrisburg, Pennsylvania.

> One congregation has taken to heart the ambitious idea of "green building." St. Stephen's Cathedral, a historic landmark in downtown Harrisburg, Pa., is undergoing a major renovation that could win it a "gold" rating from the US Green Building Council of the building industry. That rating in the LEED program (Leadership in Energy and Environmental Design) "would make us the first historic renovation project ever to get the designation," says the Rev. Malcolm McDowell, dean of St. Stephen's.[44]

There are also newly constructed places of worship that have, in varying degrees, made green statements and commitments. One example is St. Gabriel's Passionist Parish in suburban Toronto, Canada. An architectural review in the *Globe and Mail* had this analysis.

> Rather than creating an introverted experience of worship inspired by stained glass windows, the emphasis has been placed on the mystery of the natural world. Views are directed to the outdoor gardens just beyond the massive clear-glass curtain wall. . . . Oriented to the south, the church embraces the sun and disperses it throughout its spaces.
> On paper, St. Gabriel's may well become the most sustainable church in Canada. But there is another message that comes from a walk through the garden where the stations of the earth are to be located. They chart the evolution and trauma of the universe, from the big bang, to the bursting forth of flowers, to the beginning of agriculture, to another station depicting the atomic bomb's mushroom cloud. "To wantonly destroy a living species is to silence forever a Divine Voice" wrote Thomas Berry, the earth scholar.[45]

St. Gabriel's has been awarded Gold Leadership in Energy and Environmental Design status. In an important way, this project is a culmination of more than 25 years of conversations between the Passionist Community in Canada and Thomas Berry. The small religious community that financed and built this church has an average age of well over 70 years. Berry, the inspiration for the design, was born in 1914. The vibrant congregation of all ages that worships in this space has been fully part of the transformation from a 1950s energy-wasting, inward-focused space to the current structure that embodies a practice of green hope.

INSTITUTIONAL CONGREGATIONAL COVENANTS FOR ECOLOGICAL RENEWAL

Typically such dramatic renovations or new structures are not how congregations commit to a practice of green hope. The interplay of people, beliefs, and physical resources is complex. As Sherkat and Ellison point out, "Both 'religion' and the 'environment' imply an intersection of subjective understandings held by active agents, and physical resources or formalized rules or principles."[46] The conclusions reached by Sherkat and Ellison are not self-evidently accurate in today's more heated environmental climate of awareness and involvement,[47] because of the age of their data sample (1993 General Social Survey). However, the measures they studied provide a useful map of key issues: private environmental behaviors, political environmental activism, willingness to sacrifice for the environment, belief in problem seriousness, religious factors, political conservatism, and demographic factors (age, race, gender, region of country, etc.).[48]

In all these measures there is evidence in some institutions of a positive interaction of Christianity and the natural environment at a congregational level and in religious institutions. For example, "The Quaker lobby Friends Committee on National Legislation (FCNL), a partner of Greater Washington Interfaith Power and Light, recently completely rebuilt their historic building to incorporate the best green technologies. The building has a vegetated roof, geothermal heating and cooling, bamboo floors, a light 'scoop,' and accommodation for bicycle commuters."[49]

What is it about congregations (or intentional communities that have similar characteristics) that make them likely places for ecological renewal? John Cobb Jr. begins a reflection on the role of institutions with a statement that singles out that role:

> Nothing that has happened in the ensuing quarter century has changed my conviction that we are collectively moving toward destruction. Nor do I feel much better about participating in institutions, such as schools and churches that continue to be part of the problem more than part of the solution. But I have realized, more and more vividly, that simply recognizing the danger does little good. Also, changing individual lifestyles, desirable as that is, does not go far toward saving the planet. If the catastrophes that are already happening, and the greater ones that are now inevitable, are to be contained and limited, there must be profound changes in institutions.[50]

Cobb's audience for this talk (now an online article) was a seminary faculty. We take responsibility for shifting the focus to local congregations and for taking a more hopeful view than that of John Cobb in 1990 of what individual congregations can do and more importantly are doing. Not all, not the majority, but many are practicing hope. "Even within our present institutional structures a bureaucrat with vision and political skills can bring about changes that are genuinely Christian. . . . Such reform requires a depth of reflection and a breadth of participation that are difficult, but not impossible, to secure."[51]

The breadth of participation to which Cobb refers is happening on a scale that would have been unimaginable even a decade ago. A key actor in the dramatic rise in the involvement of Jewish, Christian, and Muslim congregations is the Regeneration Project, founded in 1993 by Sally Bingham, an Episcopal priest, to "deepen the interconnection" between ecology and faith. In May 2007, Jewish, Christian, and Muslim leaders signed an open letter to the U.S. President and Congress calling for immediate action on global warming, signaling a shared reverence for life. But initiatives of the Regeneration Project are much larger than this, and include the Interfaith Power and Light campaign. This began in 2001 as the California Interfaith and Light campaign—to put, as Bingham says, "faith into action."[52]

"The mission of the Regeneration Project is to deepen the connection between ecology and faith. Our Interfaith Power and Light campaign is mobilizing a religious response to global warming in congregations through the promotion of renewable energy, energy efficiency, and conservation."[53] Interfaith Power and Light has affiliates in 23 states (up from a dozen in 2003). It asks for congregational and individual covenants. The initial commitment is to do *one or more* of the action steps set out in the covenant that the congregation or individual signs. For congregations, these steps range from education on energy production and usage, to encouragement of individual covenants, to making energy efficiency improvements in congregational buildings, to supporting appropriate public policies. For individuals, covenanted commitments are to take *one or more* of the following action steps: to educate oneself on energy production and usage in relation to global warming, air quality, and environmental protection; to conduct an energy audit of one's living space; to replace incandescent light bulbs; to purchase and/or install electricity from clean renewable resources; to use the most energy efficient means possible for transportation; and to support helpful public policies.[54] Congregations covenanted with Interfaith Power and Light move well beyond earnest exhortation to effective domestic and political action—committed actions not without personal cost. These congregations demonstrate that organized religion, in its best moments, "is more concerned with the welfare of the group and the collective good than is any other institution. So, [Max Oelschlaeger] concludes, the church is our last, best chance."[55] "The metaphor of caring for creation is literally an instrument for social transformation: it is an instrument of moral and intellectual growth . . . not a theological rule but an imaginative paradigm that might prove useful for a culture undergoing ecocrisis."[56] Practical action may be less challenging than changes in practices of worship, which are heart and soul of congregational life and that shape the social imaginary of each congregation.

Religious education can play an important part in greening a social imaginary. We have not, however, listed either online or print resources for religious education for an encouragingly practical reason—namely that resources are proliferating and our references, in both print and e-formats) are likely to be quickly dated. We do point to just one source of resources that might otherwise go unnoticed: KAIROS: Canadian Ecumenical Justice Initiatives (www.united-church.ca/partners/ecumenical/kairos) provides strong resources, many of which are relevant to religious education in the area of environment/ecological justice.

RITUAL, PRAYER, AND WORSHIP

There is an almost painful sensitivity to language in worship. For example, while women's liberation movements[57] certainly caused deep anxiety in many quarters, it was the experience of at least some who wished the movements well—and who tolerated if not supported its aims and causes—that there remained an edgy discomfort with the use of feminine language for God in worship. It may also have been the experience of some that facing this discomfort was elemental in really dealing with entrenched attitudes of patriarchy and the domination of women. Worship can engage the whole person deeply and can lay bare and challenge deeply ingrained attitudes of human domination over the earth. A facet of the social imaginary comes into sharp relief.

Seven Songs of Creation is a collection of liturgies created by Norman C. Habel and by colleagues and students, women and men, from Adelaide, Australia, Oshkosh, Wisconsin, an email ecotheology group in Australia and New Zealand, Andover, Massachusetts, and elsewhere. Judging from the acknowledgements, it appears that most liturgies are the work of groups rather than individuals. The seven songs are of Sanctuary, of Earth, of Sky. Cumulatively, the effect of affirmations, prayers, rituals, creeds, confessions, Eucharists, and blessings in *Seven*

Songs of Creation is to move the worshipper who enters worship unguarded to an engagement with Earth and Earth community as God's revelation and the place of God's glory. *Seven Songs of Creation* is not unique in its invitation to celebrate and heal the earth in worship. It is, however, a strong example of the role, the demanding power, and the essential role of liturgy in changing the social imaginaries of worshipping communities. Indeed, it is a new ecological social imaginary underway.

EVANGELICAL CHRISTIAN CONGREGATIONS AND INITIATIVES

One aspect of the environmental crisis for some Christians has been the challenge to their beliefs. This has been particularly so for conservative evangelical Christians. John Jefferson Davis, writing in the *Journal of the Evangelical Theological Society*, argues, "Certain 'blind spots' in the structure and content of recent Evangelical systematic theologies have contributed to the neglect of environmental issues and environmental stewardship in certain segments of the evangelical subculture."[58] Specifically,

> Recent biblical scholarship has begun to recognize the *cosmic* impact of the atoning work of Christ, and this recognition has begun to make itself felt in evangelical environmental scholarship. There seems to be a growing recognition that texts such as Col 1:20, which state that in Christ God was pleased to reconcile to himself *all things*, whether *on earth* or in heaven, through the blood of the cross, have powerful implications for an evangelical stewardship of creation.[59]

The challenge is often experienced in a powerfully personal way, moving the person to decisions that are more conversion than simply change of mind. A PBS special by Bill Moyers, "Is God Green?"[60] "examined the increasing involvement of evangelical Christians in the environmental movement." Part of Moyers's special tells the story of the Vineyard Church in Boise, Idaho. This is "an evangelical nondenominational congregation, [where] Pastor Tri Robinson helped created a ministry called 'Let's Tend the Garden,' describing environmental stewardship as a 'biblical mandate'."[61] Robinson describes a struggle of more than six months as he prepared to preach this sermon, so much did he see it both as his vocation and a clear departure from the way the Bible had been preached in his congregation to this time. His born-again congregants experienced another conversion. Their God-fearing, Bible-preaching pastor led the way. Said one congregant, "I care about the Creator, therefore I care about the creation." And another said of Robinson, "He's a typical Republican Evangelical, except now he is green." Now, "Vineyard urges Christians to become involved in everything from recycling and energy conservation, to cleaning up highways and the Boise River, to working with the U.S. Forest Service to build and maintain hiking trails."[62]

On the national level, under the leadership of Richard Cizik, evangelical Christians—whose national association has 30 million members—are identifying environmental concerns and taking care of the earth as a moral issue connected to religious traditions. And they are acting on it. This represents a change in belief, a conversion, on global warming. Cizik, executive director of the National Association of Evangelicals, said in an interview last year that he experienced a profound "conversion" on the global warming issue in 2002 after listening to scientists at a retreat. Now an emblem for a new breed of evangelical environmentalists, he has been written about in *Vanity Fair* and *Newsweek* and has appeared in "The Great Warming," a documentary on climate change.[63]

Such was the credibility of the scientist John Houghton, and the moral authority of Richard Cizik that the power of sustained conversation (what King calls the dynamics of reciprocity that underpin supportive social groups) that "Climate Change: An Evangelical Call to Action," was approved by the National Association of Evangelicals, October 8, 2004. This represents a striking extension of moral concerns of the evangelical community's long-standing commitment to the sanctity of human life. The declaration is framed in five claims: (1) human-induced climate change is real; (2) the consequences of climate change will be significant, and will hit the poor the hardest; (3) Christian moral convictions demand our response to the climate change problem; (4) the need to act now is urgent; and (5) governments, businesses, churches, and individuals all have a role to play in addressing climate change—starting now.[64]

Theological perspectives vary widely within Christian denominations. And while the "intellectual basis of Christian concern for *nature as beloved of God* is not the same as . . . concern for the *intrinsic value of nature in and for itself,*"[65] which is the theological orientation of many Christians, the shift to a commitment to creation care by 30 million evangelical Christians is notable and important.

This is not the first initiative within evangelical Christianity to take seriously the need for responsible "creation care."[66] In the late

1970s, the Au Sable Institute (now with campuses in the Great Lakes, Pacific Rim, South Florida, India, and Africa) grew out of the Au Sable Trails Science Camp for Youth in northern Michigan, and was transformed into an educational organization of scientists who were also committed Christians. In this it is singular—this is an educational enterprise of well-credentialed scientists with a unique Christian vision. It describes the Au Sable idea this way: "The boundaries of Au Sable Institute are the boundaries of the biosphere. The reach of its service is the whole Creation. It works with others to support and restore all creatures great and small. It integrates science, ethics, and praxis using the resources of multiple disciplines. It respects Creation as both teacher and evangelist."[67]

Similar new projects are springing up in the United States. The Creation Care Study program describes itself as a "high-caliber academic semester abroad connecting Christian faith with the most complex, urgent global issues."[68] Also for evangelicals are many online networks and magazines; for those interested in the scope, shape, and concerns of Evangelical Christians for Creation (not "the environment") a good example is the Evangelical Environmental Network and *Creation Care* magazine.[69] These are but a few examples in a very long and encouraging list.

INITIATIVES OF HISTORICAL AND ECUMENICAL CHURCHES

The so-called historical and ecumenical churches (formerly in the United States and Canada, the mainline churches, now often referred to as the "oldline" churches—Episcopal, Methodist, Presbyterian, Lutheran, Congregationalist, etc.) have taken many noteworthy initiatives. For example, already noted is the Interfaith Power and Light Movement.[70] In this section we highlight some initiatives of the worldwide Anglican Communion in particular because of their ruggedly and determinedly grassroots nature. This example from the Commission for the Environment, an initiative of the Anglican Diocese of Canberra and Goulburn, gets at a local, immediate, grassroots concern that is typical.

> We and the majority of our neighbors are good farmers . . . from central west NSW. How do I feel about climate change?
> Firstly—in my face at home it is gut wrenching. My husband is the 6th generation of his family on the farm. Never has

it been like it is now. . . . There are simply no physical, financial or emotional reserves left to face the situation we are in. . . . The humiliation of dealing with the businesses in town to whom you owe money is another story.

What . . . so many urban dwellers don't realize is the spiritual connection to the land that farmers feel. Farmers, perhaps more than any other profession, are faced with God the Creator on a daily basis. As I drive through the countryside the sight of huge numbers of established eucalypts dying beside the road and in paddocks from moisture-stress is apocalyptic. These are supposed to be the superbly adapted species. As creation is groaning at present we on the land weep.[71]

On a larger scale, the grassroots character of the process of the Anglican Communion was evident in the meeting of the Anglican Communion Environmental Network (ACEN), in 2005 in Canberra, Australia, which addressed the church's response to global warming and related issues such as rising sea levels, droughts, and increased storms and floods. The report began with a theological statement and concluded with a plan for action. But in between these was an acknowledgment of the grassroots sources of precise evidence that fueled the passion of the conference.

Our delegates from Kenya and the Philippines reported an increase in the range of mosquitoes, resulting in more widespread malaria. This is attributed in part to rising temperatures. Projected further increases of 1.5°C to 5.8°C by 2100 will further exacerbate this trend.

Our delegates from Australia and Africa reported longer and more severe droughts, which, in the case of Kenya, are also coupled with uncertainty over the length and timing of the rainy season. Prolonged droughts across Africa are already affecting local food security, causing increased poverty and suffering. This trend is set to intensify under projected temperature rises.[72]

The focus on these grassroots reports as the bridge between theology and plan of action demonstrates the particular strength of a worldwide communion that has concentrated some of its efforts in an international commission on the environment. The grassroots nature of the enterprise also suggests the global character of social imaginaries at local levels involved in change toward more ecological perspectives.

ROMAN CATHOLIC INITIATIVES

We have already noted the renovation of the Motherhouse of the IHM Sisters in Monroe, Michigan. There are many other examples of green initiatives on the part of Roman Catholic nuns. *Green Sisters: A Spiritual Ecology*, by Sarah McFarland Taylor,[73] chronicles the lives of active Catholic nuns who are working to heal the earth. She writes about how these nuns have greened their vows, undertaken "ecologically sustainable living as a daily spiritual practice,"[74] and developed new forms of Western monasticism in deep continuity with older forms, in agriculture, food choices, and the greening of liturgy. It is an extraordinary story of deep connections between women, faith, ecology, and culture.

Genesis Farm describes itself as "a learning center for Earth studies, focusing on the connections between the health of our global commons of air, water, land and nature, and the health of our local communities and bioregions." The co-founder, Miriam MacGillis, is deeply rooted in a spirituality that reverences Earth as a primary revelation of the divine and is greatly influenced by the work of Thomas Berry. "Genesis Farm offers a variety of residential and non-residential programs, including an accredited masters-level course in Earth Literacy."[75] The specific content of the programs is organized primarily around the insights of ecotheologian Thomas Berry, particularly as set forth in his groundbreaking works, *The Dream of the Earth* and *The Great Work*.

Genesis Farm is local, yet it tries to influence social imaginaries widely by lifestyle and commitments, by alliances and connections, by education and theological conviction, by women's lives and those of men, and by ritual. Roger Gottlieb notes the importance of ritual. He writes,

> Genesis Farm's Miriam MacGillis has developed the "Cosmic Walk," in which participants follow a long rope in a spiral marked with significant moments in our universe's story. As in the Council of All Beings, the Cosmic Walk connects the physical body, emotional openness, and a focused awareness of the reality of our place in cosmic time.
>
> 4.5 billion years ago. Our Solar System forms from the remains of the supernova explosion. . . .
>
> 100 thousand years ago. Modern Humans emerge. Language, shamanic and goddess religions, and art become integral with human life.[76]

The earliest form of the cosmic walk was built at Holy Cross Centre in Port Burwell, Ontario, on the edge of a cliff overlooking Lake Erie. This cosmic walk was in the form of "The Stations of the Earth." Today there are many forms of the cosmic walk around the world.

On quite a different scale one finds the "Columbia River Watershed: Caring for Creation and the Common Good," an International Pastoral Letter by the Catholic Bishops of the Region (January 8, 2001). The bishops' concern is the 259,000 square miles of watershed on the 1,200 miles of the great river known as the Columbia. It begins in British Columbia in Canada, is fed in the United States by tributaries in Montana, Idaho, Washington, and Oregon, and flows to the Pacific Ocean. In their letter the bishops note, "The preservation of the Columbia Watershed's beauty and benefits requires us to enter into a gradual process of conversion and change. Our goal is to review very broadly the present situation of the watershed; to reflect on our common regional history; to imagine a viable, sustainable future for the watershed; and to seek ways to realize our vision."[77]

This statement demonstrates a deep knowledge of the particularities of the situation of the Columbia River watershed as seen from the perspective of Roman Catholic moral theology. The prescriptive part of the letter—the call to conversion and change—identifies as cardinal principle the conservation of the watershed as a common good for ethnic and racial cultures, including indigenous peoples, the poor, and all species of wildlife. It links resolution of economic and ecological issues, economic justice and environmental justice, the conservation of energy with the establishment of environmentally integrated alternative energy sources, and the integration of transportation and recreation needs with sustainable ecosystem requirements.[78]

More recently (January 25, 2009), Bishop Luc Bouchard of St. Paul, Alberta, issues a pastoral letter entitled "The Integrity of Creation and the Athabasca Oil Sands." The letter is a lengthy, nuanced, and often hard-hitting critique of the development of the oil sands in northern Alberta. The development of the oil sands has been mired in controversy since it began. It threatens to destroy large sections of the boreal forest, the nesting home of thousands of migrating North American bird species, one of the largest and most pristine of the continent's watersheds, and the health of the human communities downstream from the mines.[79] Bishop Bouchard addresses these problems from a Catholic perspective as follows:

> I am forced to conclude that the integrity of creation in the Athabasca Oil Sands is clearly being sacrificed for economic

gain. The proposed future development of the oil sands constitutes a serious moral problem. Environmentalists and members of First Nations and Metis communities who are challenging government and industry to adequately safeguard the air, water, and boreal forest eco-systems of the Athabasca oil sands region present a very strong moral argument, which I support. The present pace and scale of development in the Athabasca oil sands cannot be morally justified. Active steps to alleviate this environmental damage must be undertaken.[80]

These pastoral letters are serious attempts to reshape a social imaginary that made it thinkable and indeed desirable to damage large areas of the North American continent and to endanger its inhabitants for economic gain.

This chapter has given some examples of living *as if* or of lives underway to a new social imaginary. They have come from a variety of theological perspectives; from points of view more or less informed by ecological science, and from many places on earth. Collectively they point to a place on the horizon beyond which is a renewed social imaginary.

Conclusion

This book began in a spiral into despair. After almost a year of reading and rereading ecotheological texts, we became overwhelmed by the scope and seeming intransigence of ecological degradation and the long drift into chaos. Ours is a culture in ecological crisis. The grinding urgency of the problem and the cultural, political, and economic changes required by viable solutions stood in stark contrast to our experience of the virtual paralysis of "business as usual." Was there no way out?

The pivotal moment that began the movement from despair to the hope that generated this book happened in our reading of Taylor's *Modern Social Imaginaries*. Ecological problems seemed intractable precisely because they permeated every facet of everyday life—that we knew. But what we recognized in our reading of Taylor was that the social imaginary, the interweave of understandings and practices that have moral legitimacy within a given society, is malleable precisely because it had to do with every facet of everyday life. Simply put, whether intentionally or unintentionally we create the life we imagine together. The present we inhabit is not only a product of the past, but also of the futures we imagine. It is both how we read the past and how we imagine the future that sets the course of the social imaginary—how we live our lives together.

As we reread the texts of ecotheologians with this in mind we began to see both history and imagined futures laid out not as abstractions but as real engagement in the social imaginary. Fifty years of Christian ecotheological texts are neither utopic nor despairing; they persist in engaging the everyday realities of our lives together, locally and globally. Theologically, we read this as the intentional confrontation of decline and drift and a conscious redirection toward a more responsible ecological imaginary. Traditionally articulated doctrines of creation, sin, and redemption are reframed ecologically. They undergird an effective ecological ethic.

The impact of social location on both ecological destruction and practices of hope again grounds the quest for a redirected and ecologically sound social imaginary, which is lived in the specificity of everyday life—this place, this time, these people, this land, this economy. Over the years of its development, ecotheology has moved toward an increasing realization of the importance of context and local community. Practices of hope come to life in real people, in real places, in real communities, in real time. These are sources of global change.

Our final chapter presents practices that are examples that demonstrate how engaged texts have taken root and can be seen to redirect intentionally the social imaginary. Instances of "living *as if*" are in large part traceable to and dependent on ecotheological texts such as those we have presented.

Hope does not exist in the abstract; it exists in the concrete. For Christian communities who engage ecotheological texts, lived practices of hope influence the way human societies function in relation to the rest of the natural world.

NOTES

INTRODUCTION

1. David W. Rutledge, *Humans and the Earth: Toward a Personal Ecology*, (New York: Peter Lang, 1993), p. 4. See also his comments on the numbing effect of relevant statistics, pp. 11–12.
2. Christopher Turner, *Geography of Hope: A Tour of the World We Need* (Canada: Random House, 2007), p. 4.
3. Cf. Jürgen Moltman, *Theology of Hope: On the Ground and the Implications of a Christian Eschatology*. Translated by James W. Leitch (New York: Harper & Row, 1967); Ernst Bloch, *The Principle of Hope*, 3 vols., (Cambridge: MIT Press, 1986); Nicholas Lash, *A Matter of Hope* (Notre Dame: University of Notre Dame Press, 1982); John Macquarrie, *Christian Hope* (New York: Seabury Press, 1978); M. Douglas Meeks, *Origins of the Theology of Hope* (Philadelphia: Fortress Press, 1974); Karl Rahner, *Theological Investigations*, vol. 10 (London: Darton, Longman, and Todd, 1973), pp. 242–259, to name a few.
4. Readers who wish to explore the impact of texts of other religions in shaping social imaginaries will find an excellent starting point and resource at FORE's website, http://fore.research.yale.edu/ (accessed August 28, 2009).
5. Bill McKibben, *Hope, Human and Wild: True Stories of Living Lightly on the Earth* (St. Paul, MN: Ruminator Books, 2002), p. 3.
6. For a brief summary of recent scholarship on language see Rutledge, *Humans and the Earth*, pp. 12–17.
7. Ibid, p. 14. Reference to Howard Isham, "Wilhelm von Humboldt," in *Encyclopedia of Philosophy* (NY[sic]: Macmillan, 1967) vol. 4, p. 73.

8. Chaia Heller, *Ecology of Everyday Life: Rethinking the Desire for Nature* (Montreal, New York, London: Black Rose Books, 1999), esp. pp. 1–10.
9. Jan Zwicky, "Introduction," in *Hard Choices: Climate Change in Canada*, ed. Harold Coward and Andrew J. Weaver (Waterloo, ON: Wilfrid Laurier University Press, 2004), p. 9. Citation from Herakleitos (Diels, 1934, Fr. 55).
10. Ibid. Reference is to *Tao Te Ching*, ch. 1.
11. Anyone who has studied curriculum theory will note its influence in this construction. E.g., Wm. Pinar, W. Reynolds, Patrick Slattery, Peter Taubman, *Understanding Curriculum* (New York: Peter Lang, 2000).
12. This insight and structure comes from Charles Taylor, *Modern Social Imaginaries* (Durham, NC: Duke University Press, 2004).
13. Elizabeth Johnson, *She Who Is* (New York: Crossroad, 1992), p. 3.
14. See footnotes 4, 5, and 6.
15. Taylor, *Social Imaginaries*, p. 23.
16. Ibid., p. 23.
17. Ibid., p. 29.

CHAPTER 1.
THE SOCIAL IMAGINARY AND THE ECOLOGICAL CRISIS

1. Catherine Owen, *The Wrecks of Eden* (Toronto: Wolsak and Wynn, 2001), p. 39.
2. Thomas Berry, *The North American Continent*, an unpublished poem.
3. The answer is not really "simple." The references in this paragraph give that answer either implicitly or explicitly. This book deals with the attempts to address, primarily, the fallout of Western modernity, recognizing its positive as well as negative contribution. A consideration of factors that have ecologically negative consequences in other cultures and are not necessarily a product of Western modern developments would be a whole other project. We are aware, however, that ecological problems have roots other than modernity in all cultures. Our contention is that even those problems are exacerbated and intensified under modernity in the West and its far-reaching tentacles within virtually all other contexts.

4. Thus the name of his book, *Modern Social Imaginaries*. See also Charles Taylor, *A Secular Age* (Cambridge, MA and London: Belknap Press of Harvard University Press, 2007), esp. pp. 159–211.
5. Those cited by Taylor include Jürgen Habermas, *The Structural Transformation of the Public Sphere. An Inquiry into a Category of Bourgeois Society* [Orig.: Strukturwandel der Öffentlichkeit. Untersuchungen zu einer Kategorie der bürgerlichen Gesellschaft], Series: Studies in Contemporary German Social Thought (Cambridge, MA: MIT Press, 1991 [1962/1989]; Mircea Eliade, *The Sacred and the Profane* (New York: Harper, 1959); Ernst Gellner, *Nations and Nationalism Since 1790* (Cambridge, UK: Cambridge University Press, 1990); Francis Fukuyama, *Trust* (New York: Free Press, 1986); Michael Mann, *The Sources of Social Power* (Cambridge, UK: Cambridge University Press, 1986); Hubert Dreyfuss, *Being in the World* (Cambridge, MA: MIT Press, 1991). Many of these rely on the works of Rousseau, Heidegger, Wittgenstein, Polanyi, Arendt, and Bellah. Cf.: Taylor, *Social Imaginaries*, pp. 197–215.
6. Benedict Andersen, *Imagined Communities: Reflections on the Origin and Spread of Nationalism*, rev. ed. (London and New York: Verso, 1983).
7. Ibid., p. 6. Reference is to Ernest Gellner, *Thought and Change* (London: Weidenfeld and Nicholson, 1964), p. 169.
8. Taylor, *Social Imaginaries*, p. 2.
9. Cornelius Castoriadis, *The Imaginary Institution of Society* (Cambridge, UK: Polity Press, 1997).
10. Taylor, *Social Imaginaries*, p. 2.
11. Ibid.
12. Ibid., p. 23.
13. Ibid., p. 25.
14. Ibid.
15. Ibid., p. 26.
16. Ibid., p. 29.
17. Ibid., p. 71.
18. For example, Taylor comments that the three cultural forms "represent a penetration or transformation of the social imaginary by the Grotian-Lockean theory of moral order." *Social Imaginaries*, p. 69. The Grotian-Lockean theory emerges from other historic situations, events, and ideas, such as Christendom, the Protestant Reformation, advance of commerce, and so forth.
19. Ibid.

20. Ibid.
21. Carol Johnston, *The Wealth or Health of Nations: Transforming Capitalism* (Cleveland: Pilgrim Press, 1998).
22. Carolyn King makes this point in *Habitat of Grace: Biology, Christianity, and the Global Environmental Crisis; ATF (Australian Theological Forum)* Series: Three (Hindmarsh, Australia: Australian Theological Forum and Adelaide: Openbook Publishers, 2002), p. 99.
23. Taylor, *Social Imaginaries*, pp. 83–99.
24. Ibid., p. 85.
25. Ibid., p. 110.
26. Ibid., p. 111. See also, the link Taylor makes, however briefly, to the ways churches were governed, p. 109.
27. Ibid., p. 113.
28. Ibid.
29. Cf. James Nash, "Natural Law and Natural Rights," *Encyclopaedia of Religion and Nature,* ed. Bron Taylor, 2, (New York: Thoemmes Continuum, 2005) pp. 1169–1171. See also Joseph Boyle, "Natural Law," in *The New Dictionary of Theology*, ed. Joseph A. Komonchak, Mary Collins, and Dermot A. Lane (Wilmington, DE: Michael Glazier, 1988), pp. 703–708.
30. Cf. P. M. Heimann, "Voluntarism and Immanence; Conceptions of Nature in Eighteenth-Century Thought," in *Philosophy, Religion and Science in the 17th and 18th Centuries, Library of the History of Ideas*, vol. 2, edited by John W. Yolton (Rochester, NY, and Woodbridge, Suffolk: University of Rochester Press, 1990), pp. 393–405.
31. Taylor, *Social Imaginaries,* pp. 184–185.
32. Ibid., p. 186.
33. Ibid., p. 187.
34. Ibid., p. 193.
35. Ibid., p. 194.
36. Ibid., pp. 3–22.
37. Bruno Latour, *Politics of Nature: How to Bring the Sciences into Democracy* (Cambridge, MA: Harvard University Press, 2004).
38. Ibid., p. 65.
39. Roderick Nash, "The Greening of Religion," in *This Sacred Earth: Religion, Nature, Environment*, ed. Roger S. Gottlieb (New York: Routledge, 1996), 1st ed., p. 196.
40. Taylor, *Social Imaginaries*, p. 57. Immediate reference is to Karl Jaspers, *Vom Ursprung und Ziel der Geschicte* (Zürich: Artemis, 1949). Others include S. N. Eisenstadt (ed.), *The Origins and*

Diversity of Axial Age Civilizations (Albany, NY: State University of New York Press, 1986), and Robert Bellah, "Religious Evolution," in *Beyond Belief* (New York: Harper & Row, 1970), chap. 2. While not cited by Taylor, the contrast of sacred place and time in the religious (akin to pre-axial) and the nonreligious (modern) human is discussed extensively in Eliade, *The Sacred and the Profane*.

41. Ibid., p. 57.
42. Ibid., pp. 61–63.
43. Ibid., p. 63.
44. William Sims Bainbridge, "Computer Simulation of Cultural Drift," *Journal of the British Interplanetary Society*, vol. 37, 1984, pp. 420–429. http://mysite.verizon.net/wsbainbridge/dl/drift.htm#Part2 (accessed August 28, 2009).
45. Bernard Lonergan, *Insight: A Study in Human Understanding* (San Francisco: Harper & Row, 1978), pp. 226–232.
46. Ibid. Compare Taylor, *The Ethics of Authenticity* (Cambridge, MA: Harvard University Press, 2002), esp. pp. 109–121. Originally published in Canada as *The Malaise of Modernity*, 1991, an extended version of the 1991 Massey Lectures broadcast by CBC in its series *Ideas*, November 1991.
47. Berry, *North American Continent*.
48. McKibben, *Hope, Human and Wild*, p. 3.
49. Taylor, *The Ethics of Authenticity*, pp. 120–121.
50. Ibid.

CHAPTER 2. THE EMERGENCE OF ECOTHEOLOGY

1. There are many sources that present the deep historical background of environmental and related movements. See, for example, Anna Bramwell, *Ecology in the 20th Century: A History* (New Haven and London: Yale University Press, 1989; Derek Wall, *Green History: A Reader in Environmental Literature, Philosophy, and Politics* (London and New York: Routledge, 1994); Carolyn Merchant, *American Environmental History: An Introduction* (New York: Columbia University Press, 2007); Joseph Edward DeSteiguer, *The Origins of Modern Environmental Thought* (Tucson: University of Arizona Press, 2006); Ramachandra Guha, *Environmentalism: A Global History*, Longman World History Series, ed. Michael Adas (New York, Reading, MA, Menlo Park, CA, Harlow, UK, Don Mills, ON, Sydney, Mexico City, Madrid,

and Amsterdam: Addison Wesley Longman, 2000); Chad Gaffield and Pam Gaffield (eds.), *Consuming Canada: Readings in Environmental History* (Toronto: Copp Clark, 1995).

2. Cf. Karl Jacoby, "Conservation," in *Encyclopedia of World Environmental History*, vol. 1, ed. Shepard Krech III, J. R. McNeil, and Carolyn Merchant (New York and London: Routledge, 2004), pp. 262–268. It ought to be noted that Marsh, Pinchot, and Muir did not always agree on the best conservational practice. This is especially the case in the controversy over the damming of the Hetch Hetchy, which Pinchot supported and Muir, as a wilderness advocate, resisted.

3. Peter Gillis and Thomas R. Roach, "The Beginnings of a Movement: The Montreal Congress and its Aftermath, 1880–1896" in *Consuming Canada*, ed. Gaffield and Gaffield, pp. 131–151.

4. Patricia Jasen, *Wild Things: Nature, Culture and Tourism in Ontario, 1790–1914* (Toronto: University of Toronto Press, 1995).

5. See Anne Marie Dalton, *A Theology for the Earth: Contributions of Thomas Berry and Bernard Lonergan* (Ottawa: University of Ottawa Press, 1999).

6. It is also the case that some scholars see the emergence of environmentalism in the 1960s as a movement quite different from its predecessors because of the latter's grassroots base and broad reach for social reform. See Guha, *Environmentalism*, p. 3; Jacoby, "Conservation," p. 267; Merchant, *American Environmental History*, pp. 193–195; Samuel P. Hays, "From Conservation to Environment; Environmental Politics in the United States Since World War II," *Environmental Review* 6, 2 (Fall, 1982): 14–29. It is more likely the case that some elements of the past had more affinity to the 1960s movement. Preservationists such as John Muir, who left the Presbyterian Church to become a pantheistic mystic and who was dedicated to personal simplicity of life as part of his activism, is one example. Henry David Thoreau may be another. See some of their works; some are listed in the bibliography. In any case, the degree to which one views the development in the 1960s as continuous with conservation and other previous efforts and writings does not change the validity of arguments presented below.

7. Rex Weyler, *Greenpeace: How a Group of Ecologists, Journalists and Visionaries Changed the World* (Vancouver: Raincoast Books, 2004).

8. M. Williamson, "The Impact of Space Technology on Society," *Technology and Society at a Time of Sweeping Change.* Proceed-

ings International Symposium on Technology and Society, 1997. (Glasgow, UK: June, 1997), pp. 139–147.
9. Denis Cosgrove, "Contested Global Visions: *One-World, Whole-Earth*, and the Apollo Space Photographs," *Annals of the Association of American Geographers* 84:2 (June 1994): pp. 270–294. See also Michael Dear, "Who's Afraid of Postmodernism? Reflections on Symanski and Cosgrove," pp. 295–300, in the same issue.
10. Weyler, *Greenpeace*, esp. pp. 55–68.
11. Ibid., pp. 55–59.
12. Anna Bramwell, *Ecology in the 20th Century*, pp. 211–236.
13. Ibid., p. 214.
14. Rachel Carson, *Silent Spring* (Boston: Houghton Mifflin, 1962), p. 13.
15. Lisa Sideris, "The Ecological Body," in *Soundings* 85 1/2 (Spring–Summer 2002): 109.
16. Paul Brooks, *The House of Life: Rachel Carson at Work* (Boston: Houghton Mifflin, 1972), p. 299, quoting Charles Simmons to Carson, June 18, 1962.
17. Hampton L. Cross, review of *Silent Spring* by Rachel Carson, *College and University* (Spring, 1963), cited by Kimm Groshong, "The Noisy Response to Silent Spring; Placing Rachel Carson's Work in Context," Senior Thesis (Pomona College, April 2002), p. 46. Available online at www.sts.pomona.edu/ThesisSTS.pdf.
18. Herman J. Muller, review of *Silent Spring*, by Rachel Carson, *New York Herald Tribune* (23 September 1962), cited by Groshong, "The Noisy Response," p. 46.
19. Brooks, *The House of Life*, p. 296. Frank Graham Jr., *Since Silent Spring* (Boston: Houghton Mifflin, 1970), pp. 176–181.
20. J. R. McNeill, *Something New under the Sun: An Environmental History of the Twentieth Century World* (New York: W.W. Norton, 2000), p. 337 and references.
21. Cf. Ibid., pp. 325–326, for deeper background in the industrial and technological growth that speeded up in late nineteenth and early twentieth centuries.
22. Barry Commoner, *The Closing Circle* (New York: Knopf, 1971).
23. Ibid.
24. Douglas Smith, "United Nations Conference on the Human Environment (1972) Stockholm, Sweden," in *Environmental Encyclopedia*, ed. William P. Cunningham, et al. (Detroit: Gale Research, 1994), p. 854.
25. Commoner, *Closing Circle*, pp. 10–11.

26. Timothy Miller, "Hippies," in *Encyclopedia of Religion and Nature*, ed. Bron R. Taylor (London: Thoemmes Continuum, 2005), pp. 779–780.
27. Graham Harvey, "Paganism—Contemporary" in *Encyclopedia of Religion and Nature*, ed. Bron R. Taylor (London: Thoemmes Continuum, 2005).
28. Starhawk, *Webs of Power: Notes from the Global Uprising* (Gabriola Island, Canada: New Society Publishers, 2002).
29. Weyler, *Greenpeace*, pp. 28, 41, 63.
30. Louis Cox, *Truth Is Something That Happens*. Available at www.quakerearthcare.org/SpiritualityandEarthcare/index.htm (accessed August 28, 2009).
31. Joseph Sittler, *The Care of the Earth and Other University Sermons* (Philadelphia: Fortress Press, 1964), pp. 89, 91. His reference to the poem is "Advice to a Prophet," by Richard Wilbur. From *Advice to a Prophet and Other Poems* (New York: Harcourt, Brace & World, 1961). Reprinted with permission.
32. Bernard Lonergan, *Method in Theology* (New York: Herder & Herder, 1972).
33. Pierre Teilhard de Chardin, *The Phenomenon of Man*[sic], trans. Bernard Wall (London: William Collins; New York: Harper & Row, 1959). For a complete list of the works of Teilhard de Chardin, including correspondence, see Claude Cuénot, *Teilhard de Chardin* (Baltimore: Helicon Press, 1965), pp. 412–482.
34. While these theologies did not appear in substantial form until the late 1960s, the theologies had obvious effects in such significant events as the Second Vatican Council of the Roman Catholic Church (1962–1965) and in the 1968, Second General Conference of Latin American Bishops, Medellin, Colombia. For a full account of the emergence of these theologies, cf. Philip Berryman, *Liberation Theology: Essential Facts about the Revolutionary Movement in Latin America and Beyond* (New York: Pantheon, 1986); Rebecca S. Chopp, *The Praxis of Suffering: An Interpretation of Liberation and Political Theologies* (Maryknoll, NY: Orbis, 1986).
35. Scott Frickel and Neil Gross in "A General Theory of Scientific Intellectual Movements," *American Sociological Review* 70 (April 2005): 204–232.
36. C. A. Bowers, *The Culture of Denial: Why the Environmental Movement Needs a Strategy for Reforming Universities and Public Schools* (Albany, NY: State University of New York, 1997) explores resistance to ecology in the intellectual community and the way this resistance is tied to the abandonment of important intellectual traditions.
37. Frickel and Gross, p. 204.

38. John B. Cobb Jr., "The Greening of Theology," www.religiononline.org/showarticle.asp?title=1093 (accessed August 28, 2009).
39. Frickel and Gross, p. 206.
40. Ibid., p. 207.
41. Joseph Sittler, *Evocations of Grace: Writings on Ecology, Theology, and Ethics*, ed. Steven Bouma-Prediger and Peter Bakken (Grand Rapids, MI: Eerdmans, 2000), p. 7.
42. H. Paul Santmire, "Toward a Christology of Nature: Claiming the Legacy of Joseph Sittler and Karl Barth," *Dialog* 34 (Fall 1995): 270. Cited in Sittler, *Evocations of Grace*, p. 7.
43. Rosemary Radford Ruether, *Women Healing Earth: Third World Women on Ecology, Feminism, and Religion* (Maryknoll, NY: Orbis, 1996), p. 1.
44. Cobb, "The Greening of Theology."
45. Frickel and Gross, pp. 207–208.
46. Cobb, "The Greening of Theology."
47. Frickel and Gross, p. 209.
48. Ibid., p. 210.
49. Ibid., p. 212.
50. Ibid., p. 217.
51. Ibid.
52. www.religionandecology.org (accessed August 28, 2009).
53. Harold Coward's review of the AAR Religion and Ecology Group, November 29, 2004, is both history and evaluation. www.religionandnature.com/aar/R+E-Review2004.pdf (accessed August 28, 2009).
54. www.religionandnature.com/aar (accessed August 28, 2009).
55. Frickel and Gross, p. 221, quoting Arnold Thackray and Robert K. Merton, "On Discipline Building: The Paradoxes of George Sarton" *Isis* 63 (1972): 494.
56. Stephen Dunn, "Three Shades of Green, A Canadian Experiment," presentation given at an eco-theology conference in Chicago (March, 1995).
57. Ibid., p. 2.
58. Ibid.
59. Ibid.
60. Ibid., p. 3.
61. Ibid., pp. 2–3.
62. Ibid., p. 3.
63. Ibid., p. 212.
64. Readers who wish to pursue similar projects at an undergraduate level might read James Schaefer's, "Environmental ethics from an interdisciplinary perspective: the Marquette experience," in a

special issue on teaching environmental ethics of *Worldviews: Environment, Culture, Religion* (Brill, 8, 2–3, 336–352) 2004. More comprehensive is "Religious Studies and Environmental Concern" in the *Encyclopedia of Religion and Nature*, ed. Bron R. Taylor (New York: Thoemmes Continuum, 2005), pp. 1373–1379.
65. Ibid., p. 225.
66. Ibid., p. 225.
67. Ibid., p. 226.
68. Ibid., p. 225.
69. George Myerson, *Ecology and the End of Postmodernity* (Cambridge, UK: Icon/Totem Books, 2001), p. 70.
70. Ibid.
71. Ibid.
72. Ibid., p. 72, quoting Ulrich Beck and Elizabeth Beck-Gernsheim, *The Normal Chaos of Love*, trans. by Mark Ritter and June Wiebel (Cambridge, UK: Polity Press, 1995), p. 3.

CHAPTER 3. IMAGINED FUTURES

1. Steve Zavestoski, "Constructing and Maintaining Ecological Identities: The Strategies of Deep Ecologists," in *Identity and the Natural Environment: The Psychological Significance of Nature*, ed. Susan Clayton and Susan Opotow (Cambridge, MA: MIT Press, 2003), pp. 297–316.
2. Susan Clayton and Susan Opotow, eds., *Identity and the Natural Environment: The Psychological Significance of Nature* (Cambridge, MA: MIT Press, 2003), p. 6.
3. Zavestoski, p. 312.
4. Charles Taylor, *Modern Social Imaginaries* (Durham, NC: Duke University Press, 2004), p. 23.
5. Ibid., p. 115.
6. Willett Kempton and Dorothy C. Holland, "Identity and Sustained Environmental Practice," in *Identity and the Natural Environment*, eds. Susan Clayton and Susan Opotow (Cambridge, MA: MIT Press, 2003), pp. 317–342.
7. Cf. Brad Allenby, "The Anthropogenic Earth: Integrating and Reifying Technology, Environmentalism, and Religion," *Princeton Seminary Bulletin*, vol. 24, no. 1 (2003):104–121.
8. Anne Marie Dalton, *A Theology for the Earth* (Ottawa: University of Ottawa Press, 1999), pp. 141–149.

9. H. Paul Santmire, *Brother Earth: Nature, God and Ecology in Time of Crisis* (New York: Thomas Nelson, 1970), pp. 37–45 and 60–64.
10. Joseph Sittler, "Ecological Commitment as Theological Responsibility," *Zygon* 5 (1970): 177.
11. Rudolph Bultmann, *Jesus and the Word* (New York: Charles Scribner, 1934). Cited in Harold H. Oliver, "The Neglect and Recovery of Nature in Twentieth-Century Protestant Thought," *Journal of the American Academy of Religion*, vol. 60, no. 3 (Autumn, 1992): 382–383.
12. Peter W. Bakken, "Introduction: Nature as a Theater of Grace: The Ecological Theology of Joseph Sittler," in *Evocations of Grace: Writings on Ecology, Theology, and Ethics, Joseph Sittler*, eds. Steven Bouma-Prediger and Peter Bakken (Grand Rapids, MI: William B. Eerdmans, 2000), p. 3.
13. Ibid., p. 5.
14. Sittler. "Ecological Commitment," p. 173.
15. Joseph Sittler, "A Theology for the Earth" *The Christian Scholar*, vol. xxxvii, no. 3 (1954): pp. 367–374.
16. Ibid., p. 372.
17. Joseph Sittler, *The Care of the Earth and other University Sermons* (Philadelphia: Fortress Press, 1964), p. 98.
18. Ibid. p. 43.
19. Ibid.
20. Joseph Sittler, "Called to Unity" (1962), in *Evocations of Grace*, p. 40.
21. Joseph Sittler, "Evangelism and the Care of the Earth," (1973), in *Evocations of Grace*, p. 203.
22. Peter W. Bakken, "Introduction," in *Evocations of Grace*, p. 12.
23. Ibid., p. 13.
24. Ibid., p. 14.
25. Cf. "Secondary Literature/Applications," in *Evocations of Grace*, p. 237. A number of recent books, articles, and theses are listed.
26. Thomas Berry, *The Great Work: Our Way into the Future* (New York: Bell Tower, Random House, 1999), p. 3. See also Anne Marie Dalton, "Who Cares About the Meadow?" in *Every Grain of Sand: Canadian Perspectives on Ecology and Environment*, ed. J. A. Wainwright (Waterloo, ON: Wilfrid Laurier Press, 2004), pp. 73–86.
27. Thomas Berry, "The Dream of the Earth: Our Way into the Future," *Dream of the Earth* (San Francisco, CA: Sierra Club, 1988), p. 205.

28. Brian Swimme and Thomas Berry, *The Universe Story: From the Primordial Flaring Forth to the Ecozoic Era, A Celebration of the Universe* (New York: HarperCollins, 1992).
29. Thomas Berry, "Economics as a Religious Issue," in *Dream of the Earth*, p. 88.
30. Thomas Berry, "Contemporary Spirituality: The Journey of the Human Community," *Cross Currents* (Summer/Fall 1974): 176.
31. Thomas Berry, "Creative Energy," in *The Dream of the Earth*, p. 10. Originally presented at the Conference on the Future of India, State University of New York, 1976. For an account of the influence of Teilhard's work on Berry, see Dalton, *Theology for the Earth*, pp. 61–75.
32. Thomas Berry, "Every Being Has Rights," *Twenty-third Annual E.F. Schumacher Lectures*, ed. Hildegrade Hannum (Stockbridge, MA: E.F. Schumacher Society, 2004), pp. 13–15.
33. Sarah McFarland Taylor, *Green Sisters: A Spiritual Ecology* (Cambridge, MA: Harvard University Press, 2007), p. 5.
34. Rosemary Radford Ruether, *New Woman, New Earth: Sexist Ideologies & Human Liberation* (New York: Seabury Press, 1975), pp. 186–218.
35. Ibid.
36. Ibid., p. 186. Reference is to William Leiss, *The Domination of Nature* (New York: Braziller, 1972).
37. Ibid., p. 204.
38. Barbara Darling-Smith, "Rosemary Radford Ruether," in *Encyclopedia of Religion and Nature*, ed. Bron Taylor, et al. (New York: Thoemmes Continuum, 2005), vol. 2, pp. 1433–1434.
39. Rita M. Gross and Rosemary Radford Ruether, *Religious Feminism and the Future of the Planet: A Christian-Buddhist Conversation* (New York: Continuum International Publishing Group, 2001), p. 55.
40. Rosemary Radford Ruether, ed., *Gender, Ethnicity, and Religion: Views from the Other Side* (Minneapolis: Fortress Press, 2002), p. ix.
41. Chaia Heller, *Ecology of Everyday Life: Rethinking the Desire for Nature* (Montreal, New York, London: Black Rose Books, 1999), p. 93.
42. Rosemary Radford Ruether, *Gaia and God: An Ecofeminist Theology of Earth Healing* (San Francisco: HarperCollins, 1992), pp. 268–274.
43. Gross and Ruether, *Religious Feminism and the Future of the Planet*. See Ruether's defense of her own position, pp. 56–58.

44. Ibid. See a summary of her positions on classic doctrines and their oppressive interpretations, pp. 88–101.
45. Rosemary Radford Ruether, *Sexism and God-Talk: Toward a Feminist Theology* (Boston: Beacon Press, 1983), p. 257.
46. Ibid., p. xiv.
47. John B. Cobb Jr., "A Critical View of Inherited Theology," *Christian Century* (1980).
48. John B. Cobb Jr., *Is It Too Late? A Theology of Ecology*, revised ed. (Denton: Environmental Ethics Books, 1995).
49. Cobb, "A Critical View of Inherited Theology," p. 194.
50. Ibid., p. 196.
51. John B. Cobb Jr., *Reclaiming the Church*.
52. Herman E. Daly and John B. Cobb Jr., *For the Common Good: Redirecting the Economy toward Community, the Environment, and a Sustainable Future* (Boston: Beacon Press, 1989).
53. John B. Cobb Jr., *Sustainability: Economics, Ecology and Justice* (Eugene, OR: Wipf and Stock Publishers, 2007).
54. Charles Birch and John B. Cobb Jr., *The Liberation of Life: From the Cell to the Community* (Cambridge, MA: Cambridge University Press, 1981).
55. Christopher Ives and John B. Cobb, Jr. eds., *The Emptying God: A Buddhist-Jewish-Christian Conversation* (Eugene, OR: Wipf and Stock, 1990).
56. John B. Cobb Jr., "Christianity, Economics, and Ecology," in *Christianity and Ecology*, p. 498.
57. Ibid., p. 526.

CHAPTER 4. THEOLOGY AND THE ECOLOGICAL CRISIS

1. Bernard Lonergan, *Insight: A Study of Human Understanding* (New York, Hagerstown, San Francisco, London: Harper & Row, 1978), pp. 226–232.
2. We are grateful to Stephen Dunn for alerting us to this comparison. See Lonergan, *Insight*, pp. 247–250.
3. Cf. Kathryn Tanner, *Theories of Culture: A New Agenda for Theology*. Guides to Theological Inquiry Series (Minneapolis: Fortress Press, 1997).
4. Ibid. Tanner distinguishes academic theology from a more general theology in which most if not all believers participate. One of the distinguishing features of academic theology, which she sees as a

social practice in its own right, is the accessibility to such resources. See, pp. 80–82.
5. Ibid., p. 80.
6. Cf. Tanner, *Theories of Culture*, pp. 138–151, and ch. 7, "Diversity and Creativity in Theological Judgment," pp. 156–175. See also, Paul Lakeland, *Postmodernity: Christian Identity in a Fragmented Age: Guides to Theological Inquiry* (Minneapolis: Fortress Press, 1997), esp. pp. 108–113, for example, for a proposal regarding the interpretation of the uniqueness of Christ within a radical postmodernism.
7. Tanner, *Theories of Culture*, pp. 72–80.
8. Thomas Berry, *The Dream of the Earth* (San Francisco: Sierra Club Books, 1988), pp. 124–130, Critiques are scattered throughout all his works. For a synthetic overview see Dalton, *A Theology for the Earth*, pp. 111–118. Carolyn Merchant, *The Nature of Death: Women, Ecology and the Scientific Revolution* (New York: Harper & Row, 1989), and *Reinventing Eden* (New York and London: Routledge, 2003).
9. Giovanna Di Chiro, "Nature as Community: The Convergence of Environment and Social Justice," in *Privatizing Nature: Political Struggles for the Global Commons*, ed. Michael Goldman (New Brunswick, NJ: Rutgers University Press, 1998), p. 120.
10. Ibid., p. 121.
11. Anne Lonergan and Caroline Richards, (eds.), *Thomas Berry and the New Cosmology* (Mystic, CT: Twenty-Third Publications, 1990).
12. Eleanor Rae, comment on the back cover of Leonardo Boff, *Ecology and Liberation: A New Paradigm*, trans. John Cumming (Maryknoll, NY: Orbis, 1995).
13. Di Chiro, "Nature as Community," p. 126.
14. Report of the United Church of Christ's Commission for Racial Justice, 1987, cited in Laura Westra and Bill E. Lawson, "Introduction," *Faces of Environmental Racism: Confronting Issues of Global Justice*, 2nd ed., ed. Laura Westra and Bill E. Lawson (Lanham, MD: Rowman & Littlefield, 2001), p. xviii.
15. Ibid.
16. *West Africa*, (20 June 1988), 1107, cited in Segun Gbadegesin, "Multinational Corporations, Developed Nations, and Environmental Racism: Toxic Waste, Exploration, and Eco-catastrophe," in *Faces of Environmental Racism*, pp. 187–191.
17. Gbadegesin, "Multinational Corporations," pp. 190–198.
18. Di Chiro, "Nature as Community," p. 124.

19. Ibid., 125.
20. Ian G. Barbour ed., *Earth Might Be Fair: Reflections on Ethics, Religion, and Ecology* (Englewood Cliffs, NJ: Prentice-Hall, 1972).
21. Richard A. Young, *Healing the Earth: A Theocentric Perspective of Environmental Problems and Their Solutions* (Nashville, TN: Broadman & Holman, 1994), p. 274.
22. Catherine Keller, "Lost Fragrance: Protestantism and the Nature of What Matters," *Visions of a New Earth: Religious Perspectives on Population, Consumption, and Ecology*, edited by Harold Coward and Daniel C. Maguire (Albany, NY: State University of New York Press, 2000), pp. 79–93.
23. Catherine Keller, *Face of the Deep* with reference to Edward W. Said, *Beginnings: Intention and Method* (New York: Columbia University Press, 1985), p. 372 ff.
24. Norman Habel ed., *The Earth Bible* (Cleveland, OH: Pilgrim Press, 2001).
25. Sallie McFague, *Abundant Life: Rethinking Theology and Economy for a Planet in Peril* (Minneapolis: Fortress Press, 2001), p. 43.
26. Ibid.
27. Ibid.
28. Ibid., p. 77.
29. Cf. references in the bibliography to Thomas Berry, Herman Daly, and John Cobb Jr., Rosemary Radford Ruether, Heather Eaton and Lois Lorentzen, Larry Rasmussen.
30. Ruether, *Integrating Ecofeminism, Globalization and World Religions* (Lanham, MD: Rowman & Littlefield, 2005), pp. 1–44.
31. Ibid., p. 1.
32. Ibid., p. 3.
33. Ibid., pp. 4–5.
34. Ibid., pp. 11–33.
35. Ibid., p. 20. Reference to Vandana Shiva, *Stolen Harvest: the Hijacking of the Global Food Supply* (Boston: South End Press, 1999), and *Biopiracy: The Plunder of Nature and Knowledge* (Boston: South End Press, 1997).
36. Ibid., p. 21.
37. Ibid., p. 22.
38. Cf., for example, Heather Eaton and Lois Lorentzen, *Ecofeminism and Globalization*; James P. Martin Schramm, "Population-Consumption Issues: The State of the Debate in Christian Ethics," in *Theology for Earth Community: A Field Guide*, edited by Dieter T. Hessel (Maryknoll, NY: Orbis Press and World Council of Churches Publications, 1994).

39. Norman Habel, "Introducing the Earth Bible," *Readings from the Perspective of the Earth, Earth Bible 1*, ed. Norman Habel (Cleveland, OH: Pilgrim Press, 2000), p. 28.
40. See use of Lonergan's theology in discussion of the ecological crisis in Anne Marie Dalton, *A Theology for the Earth: The Contributions of Thomas Berry and Bernard Lonergan* (Ottawa: University of Ottawa Press, 1999).
41. Dieter T. Hessel, "U.S. Churches in the Environmental Movement," *Theology for the Earth Community: A Field Guide*, ed. Dieter T. Hessel (Maryknoll, NY: Orbis, 1996), pp. 199–207.
42. Mary Evelyn Tucker and John Grim, "Series Forward," *Christianity and Ecology*, p. xxiii.
43. Elizabeth Johnson, "Losing and Finding Creation in the Christian Tradition," *Christianity and Ecology*, pp. 3–21. Sallie McFague, "An Ecological Christology," *Christianity and Ecology*, pp. 29–45; Mark Wallace, "The Wounded Spirit as the Basis of Hope," *Christianity and Ecology*, pp. 51–72.
44. Gordon Kaufman, "Response to Elizabeth Johnson," *Christianity and Ecology*, p. 27.
45. Kwok Pui-lan, "Response to Sallie McFague," *Christianity and Ecology*, p. 50.
46. Eleanor Rae, "Response to Mark Wallace," *Christianity and Ecology*, p. 76.
47. David G. Hallman, "Climate Change: Ethics, Justice and Sustainable Community," *Christianity and Ecology*, p. 452.
48. Ibid., 467.
49. Larry Rasmussen, "Global Eco-Justice: The Church's Eco-mission in Urban Society," *Christianity and Ecology*, p. 526.
50. John Chryssavgis, "The World of the Icon and Creation," *Christianity and Ecology*, p. 94.
51. Mary Oliver, "Have you ever tried to enter the long black branches?" in *West Wind: Poems and Prose Poems* (Boston: Houghton Mifflin, 1997), p. 61. Cited in Douglas Burton-Christie, "Words Beneath the Water: Logos, Cosmos, and the Spirit of Place," *Christianity and Ecology*, p. 329.
52. Burton-Christie, "Words Beneath the Water," p. 329.
53. Ibid., p. 334.

CHAPTER 5. SCIENCE AND ECOLOGY

1. Catherine Keller and Laurel Kearns, "Introduction: Grounding Theory—Earth in Religion and Philosophy" in *Ecospirit: Religions and Philosophies for the Earth*, ed. Catherine Keller and Laurel

Kearns (New York: Fordham University Press, 2007), esp. pp. 13–17.
2. Taylor, *Social Imaginaries*, p. 2.
3. Berry, *Dream of the Earth*, p. 88. See Dalton, *A Theology of the Earth*, for a summary exposition of Berry's new cosmology and his understanding and use of myth and story.
4. Alfred North Whitehead, *Science and the Modern World* (New York: Macmillan, 1926), p. 286. Cited in Harold K. Schilling, "The Whole Earth is the Lord's" in *Earth Might Be Fair: Reflections on Ethics, Religion, and Ecology*, ed. Ian Barbour (Englewood Cliffs, NJ: Prentice-Hall, 1972), p. 118.
5. William Pollard, "The Uniqueness of the Earth," in *Earth Might Be Fair*, pp. 82–99.
6. Rosemary Radford Ruether, *Gaia and God: An Ecofeminist Theology of Earth Healing* (New York: HarperCollins, 1992), p. 40.
7. Ibid., p. 58.
8. Ibid., p. 55.
9. Ibid., pp. 140–141.
10. Ibid., pp. 250–251.
11. Ibid., p. 250.
12. Anne Primavesi, *Sacred Gaia: Holistic Theology and Earth System Science* (London: Routledge, 2000), p. xiii.
13. A colleague, John Carroll, noted that the term in Greek means "self-creating" but not "self-interpreting."
14. Primavesi, *Sacred Gaia*, pp. 1–14, including references.
15. Ibid., p. xix.
16. Ibid., esp. p. 51.
17. Ibid., p. 169.
18. Ibid., p. 179.
19. Anne Primavesi, *Gaia's Gift: Earth, Ourselves, and God after Copernicus* (London: Routledge, 2003).
20. Ibid., p. 58.
21. Ibid., pp. 118–119.
22. Anne Primavesi, "The Preoriginal Gift—and Our Response to It," in *Ecospirit*, p. 222.
23. John B. Cobb Jr. "The Making of an Earthist Christian," in *Encyclopedia of Religion and Nature*, vol. 1, p. 394.
24. Ibid., p. 395.
25. Cf. John B. Cobb Jr. and Charles Birch, *The Liberation of Life: From the Cell to the Community* (London: Cambridge University Press, 1981).
26. Cf. Jay McDaniel, "Christianity (7f)—Process Theology," in *Encyclopedia of Religion and Nature*, vol. 1, pp. 364–366.

27. Ibid., p. 26.
28. John Haught, *The Promise of Nature: Ecology and Cosmic Purpose* (Mahwah, NJ: Paulist Press, 1993), p. 17. See also his first book, *The Cosmic Adventure: Science, Religion, & the Quest for Purpose* (New York/Ramsey: Paulist Press, 1984) and *Deeper Than Darwin: The Prospect for Religion in the Age of Evolution* (Boulder, CO: Westview Press, 2003).
29. John Haught, *Is Nature Enough? Meaning and Truth in the Age of Science* (Cambridge: Cambridge University Press, 2006), esp. pp. 31–36; 140, with reference to Bernard Lonergan, "Cognitional Structure," *Collected Works*, ed. F. E. Crowe (New York: Herder & Herder, 1967), pp. 221–239.
30. Sallie McFague, *The Body of God: An Ecological Theology* (Philadelphia: Fortress Press, 1993), p. 21.
31. Ibid., pp. ix–x.
32. Ibid., pp. 38–55.
33. Ibid., pp. 104–105.
34. Sallie McFague, *Life Abundant: Rethinking Theology and Economy for a Planet in Peril* (Minneapolis: Fortress Press, 2000), p. 100. Italics original.
35. Ivone Gebara, *Longing for Running Water: Ecofeminism and Liberation* (Minneapolis: Fortress Press, 1999), p. 53.
36. Larry Rasmussen, *Earth Community, Earth Ethics* (Geneva: WCC Publications, 1996), p. 99.
37. Dieter Hessel and Larry Rasmussen, editors, *Earth Habitat: Eco-Injustice and the Church's Response* (Minneapolis: Fortress Press, 2001), pp. xi, 21.
38. James A. Nash, *Loving Nature: Ecological Integrity and Christian Responsibility* (Nashville: Abington Press; Washington, DC: The Churches' Center for Theology and Public Policy, 1991) pp. 100–102. This notion of relationality is echoed in Elizabeth Johnson's reference to feminist ethics. Cf. Elizabeth Johnson, *She Who Is: The Mystery of God in Feminist Theological Discourse* (New York: Crossroad, 1992), p. 69.
39. Denis Edwards, *Ecology at the Heart of Faith: The Change of Heart that Leads to a New Way of Living on Earth* (Maryknoll, NY: Orbis, 2007).
40. Lisa Sideris, *Environmental Ethics, Ecological Theology and Natural Selection* (New York: Columbia University Press, 2003), esp. pp. 45–90.
41. Ibid., p. 90.
42. Ibid., esp. p. 120.
43. Ibid., p. 266.

44. Cf. Kevin O'Brien, "Toward an Ethic of Biodiversity: Science and Theology in the Environmentalist Debate," in *Ecospirit*, pp. 178–195.
45. Sideris, *Environmental Ethics*, p. 202.
46. Carolyn King, *Habitat of Grace: Biology, Christianity, and the Global Environmental Crisis* (Hindmarsh, Australia: Australian Theological Forum and Adelaide: Openbook, 2002), p. 40.
47. King, pp. 204–205.
48. Ibid., pp. 87–88, including references.
49. Ibid. King gives a more detailed account.
50. Ibid., p. 88.
51. Ibid., esp. pp. 155–180.
52. Ibid., p. 160.
53. There are many theologians who address other concerns arising from bioethical research and development.
54. Celia Deane-Drummond, *Genetics and Christian Ethics* (Cambridge, UK: Cambridge University Press), p. 222. Note other works by Deane-Drummond cited in our bibliography. See also Conrad Brunk and Harold Coward, (eds.). *Acceptable Genes? Religious Traditions and Genetically Modified Foods* (New York: State University of New York, 2009).
55. Ibid., pp. 222–223.
56. Ibid., p. 244.
57. Calvin DeWitt, "The Scientist and the Shepherd," in *The Oxford Handbook of Religion and Ecology*, ed. Roger Gottlieb (New York: Oxford University Press, 2006), p. 585, fn. 1.
58. DeWitt, p. 573.
59. www.ausable.org (accessed August 28, 2009).
60. Biographical information to Brad Allenby, "The Anthropogenic Earth: Integrating and Reifying Technology, Environmentalism, and Religion," *Princeton Seminary Bulletin* 24:1 (2003), pp. 104–121.
61. Ibid., p. 105.
62. Ibid., p. 106.
63. Ibid., p. 108.
64. Ibid., p. 115.
65. Ibid., p. 121.

CHAPTER 6.
GLOBAL AND LOCAL IN THE SOCIAL IMAGINARY

1. Vernice Miller-Travis, "Social Transformation through Environmental Justice," *Christianity and Ecology*, p. 570.

2. Ada María Isasi-Díaz, *Mujerista Theology* (Maryknoll, NY: Orbis, 1996), pp. 1–2.
3. Cf. Mercy Amba Oduyoye, *Introducing African Women's Theology* (Sheffield: Sheffield Academic Press, 2001), p. 38; Chung Hyun Kyung, "Ecology, Feminism and African and Asian Spirituality: Toward a Spirituality of Eco-feminism," in *Ecotheology: Voices from South and North*, ed. David G. Hallman (Maryknoll, NY: Orbis, 1994), p. 177.
4. Miller-Travis, "Social Transformation," pp. 570–571.
5. See also Kimberley Whitney, "Greening by Place: Sustaining Cultures, Ecologies, Communities," *The Journal of Women and Religion,* 19/20 (2003): 11–25.
6. Anne Cameron, "First Mother and the Rainbow Children," *Healing the Wounds: The Promise of Ecofeminism*, ed. Judith Plant (Philadelphia: New Society Press, 1993), p. 64. Cited by Janis Birkeland, "Ecofeminism: Linking Theory and Practice," *Ecofeminism: Women, Animals, Nature*, ed. Greta Gaard (Philadelphia: Temple University Press, 1993), p. 18. See also Ariel Salleh, *Ecofeminism as Politics: Nature, Marx, and the Postmodern* (London: Zed Books, 1997), p. 17.
7. Birkeland, "Ecofeminism," p. 18.
8. For a summary and analysis of many of the issues in conflict with Christian feminist perspectives in the Roman Catholic church, cf. *Globalization, Gender and Religion: The Politics of Women's Rights in Catholic and Muslim Contexts*, ed. Jane H. Bayes and Nayereh Tohidi (New York: Palgrave, 2001), chapters 2–6.
9. Rosemary Radford Ruether, "New Woman and New Earth: Women, Ecology and Social Revolution," *New Woman, New Earth: Sexist Ideologies and Human Liberation* (New York: Seabury Press, 1983), p. 186.
10. For a challenge to some of Merchant's historiography, see Anna Bramwell, *Ecology in the 20th Century: A History* (New Haven: Yale University Press, 1989), pp. 27–28.
11. Jane Caputi, "Nuclear Power and the Sacred—or Why a Beautiful Woman Is Like a Nuclear Plant," in *Ecofeminism and the Sacred*, ed. Carol J. Adams (New York: Continuum, 1993), pp. 229–250.
12. Elizabeth Dodson Gray, *Adam's World*, video, National Film Board of Canada, 1989.
13. Karen J. Warren, *Ecofeminist Philosophy: A Western Perspective on What It Is and Why It Matters* (Lanham, MD: Rowman & Littlefield, 2000), p. 207.

14. The issue of essentialism has plagued discussions of ecofeminism. Cf. Mary Mellor, "Gender and the Environment," in *Ecofeminism and Globalization*, pp. 11–22 for a recent summary of the issue.
15. See Anne Marie Dalton, "Gender and Ecofeminism: Religious Reflections on a Case Study in Soc Son, Vietnam," *Ecotheology* 11.4 (December 2006): pp. 398–414. See also, Noël Sturgeon, "Ecofeminist Natures and Transnational Environmental Policies," in Eaton and Lorentzen, *Ecofeminism and Globalization*, pp. 91–122.
16. Ruether, *New Women, New Earth*, p. 204.
17. Ruether, *Women Healing Earth* and *Gender, Ethnicity, and Religion*, and *Integrating Ecofeminism, Globalization and Religion*.
18. Rita Lester, "The Nature of Nature: Ecofeminism and Environmental Racism in America," *Gender, Ethnicity & Religion*, ed. Rosemary Radford Ruether (Minneapolis: Fortress Press, 2002), pp. 230–246.
19. Patricia-Anne Johnson, "Womanist Theology as Counter-Narrative," in *Gender, Ethnicity, and Religion*, pp. 197–214. See also her reference to Emilie M. Townes, "Womanist Theology: Dancing With A Twisted Hip," in *Introduction to Christian Theology: Contemporary North American Perspectives*, ed. Roger A. Badham (Louisville, KY: Westminster John Knox, 1998), p. 215.
20. See other selections in *Gender, Ethnicity, and Religion*.
21. Ernst Conradie and Julia Martin, "Gender, Religion and the Environment: A University of the Western Cape Case Study," *Ecotheology* 11.4 (December 2006): 431–444.
22. Ibid., p. 441.
23. bell hooks, *Ain't I a Woman? Black Women and Feminism* (Boston: South End Press, 1981).
24. Ibid., p. 131.
25. Ibid. Cf. "Sexism and the Black Female Slave Experience," pp. 15–50, and "The Imperialism of Patriarchy," pp. 87–118, as well as comments throughout the book. See a similar account but in a Latin American context, in Maria Pilar Aquino, *Our Cry for Life: Feminist Theology from Latin America* (Maryknoll, NY: Orbis, 1994), pp. 13–18.
26. Shamara Shantu Riley, in *Ecofeminism and the Sacred*, p. 197.
27. Chung Hyun Kyung, *Struggle to be Sun Again: Introducing Asian Women's Theology* (Maryknoll, NY: Orbis, 1994), p. 95.
28. Ibid.
29. Ibid., p. 96. Similar conflations of Goddess figures and images with Mary are well known in other cultures as well.

30. For a more detailed account of the work of *Con-spirando* and a list of many of its contributors, see Judith Ress, "The Con-spirando Women's Collective: Globalization from Below?" in *Ecofeminism and Globalization: Exploring Culture, Context, and Religion* (Lanham, MD: Rowman Littlefield, 2003), pp. 147–161; and Josefina Hurtado and Ute Seibert, "*Conspirando*: Women 'Breathing Together'," *Ecumenical Review* 53.1 (2001): 90–93.
31. Ress, "The Con-spirando Women's Collective," p. 158.
32. Ibid., pp. 158–159.
33. See description of the various contexts of women involved in worship by the collective reported by Hurtado and Seibert, "*Con-spirando*," p. 90.
34. Ress, "The Con-spirando Women's Collective," p. 159.
35. Ivone Gebara, "Ecofeminism: A Latin American Perspective," *Cross Currents* 53.1 (2003): p. 102.
36. Ress, *Ecofeminism in Latin America*, pp. x–xi.
37. Cf. Heather Eaton's comprehensive account of the contemporary nature of ecofeminist theology. Heather Eaton, *Introducing Ecofeminist Theologies* (London: T & T Clark, 2005).
38. Dorothy Soelle with Shirley A. Cloyes, *To Live and to Work: A Theology of Creation* (Philadelphia: Fortress Press, 1984), esp. pp. 7–20.
39. Ibid., p. 29.
40. Ibid. This is one of the major ideas throughout her book.
41. Aruna Gnanadason, "Yes, Creator God, Transform the Earth! The Earth as God's Body in an Age of Environmental Violence," *Union Seminary Quarterly Review* 58, 1–2 (April 2005): p. 164.
42. Ibid.
43. Douglas John Hall, *Professing the Faith: Christian Theology in a North American Context* (Minneapolis: Fortress Press, 1996), p. 309. Cited in Gnanadason, "Yes, Creator God," p. 168.
44. Gnanadason, "Yes, Creator God," p. 168. Reference to Sallie McFague, *Super Natural Christians: How We Should Love Nature* (Minneapolis: Fortress Press, 1997), pp. 7–9.
45. Aquino, *Our Cry for Life*, p. 179. This phenomenon has been observed throughout the world. Cf. studies of devotion by women to Kuan Yin in China and other parts of Asia.
46. Cf. Eaton and Lorentzen, *Ecofeminism and Globalization*, esp. Heather Eaton, "Can Ecofeminism Withstand Corporate Globalization?" pp. 23–37; Ruether, *Integrating Ecofeminism, Globalization and World Religions*, esp. pp. 1–44; and Lee Hong Jung, "Healing and Reconciliation as the Basis for the Sustainability of Life: An Ecological Plea for a 'Deep' Healing and Reconciliation,"

International Review of Mission 94, 372 (January 2005): pp. 85–102.
47. Kimberley Whitney, "Greening by Place: Sustaining Cultures, Ecologies, Communities," *The Journal of Women and Religion* 19/20 (2003): p. 12.
48. Ibid.
49. Ibid. Italics original.
50. Ibid. See also, Michael Northcott, "From Environmental Utopianism to Parochial Ecology: Communities of Place and the Politics of Sustainability," *Ecotheology* 8 (2000): 71–85. For particular examples such as national and corporate power influencing Rio Earth Summit and Kyoto Climate Change Conference, see page 72.
51. Ibid. See also, Wolfgang Sachs, *Global Ecology: A New Arena of Political Conflict* (London: Zed Books, 1993).
52. Whitney, "Greening by Place," p. 15.
53. Adrian Ivakhiv, "Religion, Nature and Culture: Theorizing the Field," *Journal for the Study of Religion, Nature and Culture* 1.1 (2007): 52.
54. Pat Kerans and John Kearney, *Turning the World Right-side Up: Science, Community and Democracy* (Halifax, Canada: Fernwood, 2006), pp. 142–145.
55. Ibid. This is their main argument and is written for the Canadian context.
56. Ibid. See chapter 4.
57. Gnanadason, *Listen to the Women! Listen to the Earth!* (Geneva: WCC Publications, 2005), p. 104.
58. Eaton and Lorentzen, *Ecofeminism and Globalization: Exploring Culture, Context and Religion* (Lanham, MD: Rowman & Littlefield, 2003), p. 1.
59. Heather Eaton, *Introducing Ecofeminist Theologies,* p. 92.
60. Michael S. Northcott, *The Environment and Christian Ethics* (New York: Cambridge University Press, 1996), p. 323.
61. Northcott, "From Environmental Utopianism to Parochial Ecology," *Ecotheology* 8 (2000): pp. 71–85.
62. Ibid. p. 83.
63. Ibid., pp. 84–85.
64. Larry Rasmussen, "Global Eco-Justice: The Church's Mission in Urban Society," *Christianity and Ecology,* p. 525.
65. Peggy M. Shepard, "Issues of Community Empowerment," in *Earth Habitat: Eco-Injustice and the Church's Response,* ed. Dieter Hessel and Larry Rasmussen (Minneapolis: Fortress Press, 2001), p. 169.

66. Anne Primavesi, *Making God Laugh: Human Arrogance and Ecological Humility* (Santa Rosa, CA: Polebridge Press, 2004), p. 118.
67. Nancy Scheper-Hughes, *Death Without Weeping: The Violence of Everyday Life*, cited in Bill McKibben, *Hope, Human and Wild: True Stories of Living Lightly on the Earth* (St. Paul, MN: Ruminator, 1995), p. 123.
68. Cynthia Moe-Lobeda, *Healing a Broken World: Globalization and God* (Minneapolis: Fortress Press, 2002), pp. 157–158.

CHAPTER 7. LIVING *AS IF*

1. Anne Primavesi, *Gaia's Gift* (New York: Routledge, 2003), p. 70.
2. Our data are drawn only from churches in the United States.
3. Carolyn King, *Habitat of Grace*, p. 102.
4. Ibid., p. 94.
5. Ibid., p. 95.
6. Ibid., pp. 96–97.
7. Ibid., p. 123.
8. Ibid., pp. 106–107.
9. Ibid., p. 109.
10. Ibid., p. 110.
11. Ibid., p. 114.
12. Ibid.
13. Ibid., p. 118.
14. Ibid., p. 121.
15. We limit our personal observation to the United States and Canada. It is of congregations (primarily Protestant and Roman Catholic) in these countries that we have firsthand experience. The reader will also note that most of our text and web resources as well as physical entities to which we refer in this chapter are based the United States and Canada, the United Kingdom, and Australia and New Zealand. This is a small and nonrepresentative sample of the congregations of the world. Like the literature of the earlier chapters these congregations may, however, carry disproportionate weight in the way in which the larger world is seen and functions.
16. Carolyn King, *Habitat of Grace*, p. 121.
17. Ibid., pp. 122–123.
18. Theodore W. Johnson, "Current Thinking on Size Transition," *Size Transitions in Congregations*, edited by Beth Ann Gaede (Washington, DC: Alban Institute, 2001), pp. 3–30.

19. John Cobb Jr., "The Road to Sustainability: Progress and Regress," http://religion-online.org/showarticle.asp?title=3375 (accessed August 28, 2009).
20. Romano Guardini (n.d.), www.jknirp.com/guard18.htm (accessed August 28, 2009).
21. Carolyn M. King, "Ecotheology: A Marriage between Secular Ecological Science and Rational, Compassionate Faith," *Ecotheology* 10 (2001), pp. 40–69. Quote is from page 42.
22. William (Beau) Weston, "Gruntled Center: Faith and Family for Centrists, Sociobiology 4: The Grandmother Hypothesis," http://gruntledcenter.blogspot.com/2006/02/sociobiology-4-grandmother-hypothesis.html (accessed August 28, 2009).
23. Nicholas G. Blurton Jones, Kristen Hawkes, James F. O'Connell, "Antiquity of Postreproductive Life: Are There Modern Impacts on Hunter-Gatherer Postreproductive Life Spans?" *American Journal of Human Biology* 14 (2002), p. 185.
24. We are aware that this runs counter to the argument that protection of the young is for the sake of the preservation of an individual's genes. It does not violate King's argument that we understand that "culture is an elaboration of biology" and that we recognize that "free will must be exercised within biological constraints" (*Habitat of Grace*, p. 102).
25. www.aka-alias.net/2004/09/grandmother-hypothesis.html (accessed August 28, 2009).
26. M. Lahdenper, V. Lummaa, and A. F. Russell, "Menopause: Why Does Fertility End Before Life?" (n.d.), www.redorbit.com/news/display/?id=131393 (accessed August 28, 2009).
27. Brian Gratton and Carole Haber, "Three Phases in the History of American Grandparents: Authority, Burden, Companion," *Generations* 20:1 (Spring 1996), pp. 7–12.
28. David Biello, "The Trouble with Men," *Scientific American*, October 2007, pp. 106–107.
29. U.S. Department of Commerce, Economics and Statistics Administration, U.S. Census Bureau, 1–10, www.census.gov/prod/2003 pubs/c2kbr-31.pdf (accessed August 28, 2009).
30. Ibid., p. 1.
31. Ibid., p. 7.
32. Ibid., p. 9.
33. James T. Sykes, "Three Takes on Late-Life Crises," *The Gerontologist* 42:1 (February 2002): 140.
34. http://factfinder.census.gov/servlet/ThematicMapFrameset Servlet?_bm=y&-geo_id=01000US&-tm_name=ACS_2005_

EST_G00_M00645&-ds_name=ACS_2005_EST_G00_&-_MapEvent=displayBy&-_dBy=040 (accessed August 28, 2009).
35. Mary Anderson Cooper, "Welfare's Role in Creating a Society for All Ages: Recognizing the multigenerational impact of welfare reform," *Church & Society* (January/February 1999), p. 59.
36. Arthur Kornhaber, *Contemporary Grandparenting* (Thousand Oaks, CA: Sage, 1996), p. 192.
37. The Association of Baltimore Area Grantmakers, "Baby Boomers Give More Than Older Americans," *The Chronicle of Philanthropy*, September 1, 2005. Based on an online survey of 2,333 adults with responses weighted to reflect the key demographics of the American population.
38. Census Brief 2000, October 2003, 7.
39. W. Kip Viscusi and Joni Hersch, "The Generational Divide in support for Climate Change Policies: European Evident," Harvard Law and Economics Discussion Paper No. 504 (February 2005), http://ssrn.com/abstract=670323 (accessed May 9, 2008).
40. The Regeneration Project, Interfaith Power and Light, www.theregenerationproject.org (accessed August 28, 2009).
41. Mark Chaves, *Congregations in America* (Cambridge, MA: Harvard University Press, 2004), p. 221.
42. The latest U.S. Congregational Life Survey Reports, from the Hartford Institute for Religion Research, http://hirr.hartsem.edu (accessed August 28, 2009).
43. www.ihmsisters.org/www/Sustainable_Community/Sustainable_Renovation/sustainmep.asp (accessed August 28, 2009).
44. Jane Lampman, "Churches Go Green," *Christian Science Monitor*, October 23, 2003, www.csmonitor.com/2003/0123/p11s02-lire.html (accessed August 28, 2009).
45. Lisa Rochon, *The Globe and Mail*, Cityspace, "St. Gabriel's Church: Cityspace, Seeing the Light on Sheppard," October 2, 2006: R4.
46. Darren Sherkat and Christopher Ellison, "Structuring the Religion-Environment Connection: Identifying Religious Influences on Environmental Concern and Activism," *Journal for the Scientific Study of Religion* (2007) 46(1): 71.
47. A report on research by Bernadette Hayes and Manussos Marangudakis, "Religion and environmental issues within Anglo-American democracies," *Review of Religious Research*, 2000, 42:2, 159–174, is also somewhat dated by its use of the International Social Surgery Program's Environment survey. However, one of its major conclusions, namely that Christians and non-Christians do not sig-

nificantly differ regarding their analysis does demonstrate conclusively that Lynn White's thesis that the historic root of the current ecological crisis originates in the Book of Genesis is not sustainable (p. 159).
48. Ibid., pp. 75–76.
49. www.fcnl.org/press/releases/globalwarm092606.htm (accessed August 28, 2009).
50. John Cobb Jr., "The Challenge to Theological Education," 1990, www.religion-online.org/showarticle.asp?title=1485 (accessed August 28, 2009).
51. Ibid.
52. In 1993, Bingham founded the Regeneration Project, a nonprofit organization that created a national Interfaith Power and Light Campaign to unite all faiths in efforts to prevent the catastrophic effects of global warming. www.purposeprize.org/finalists/finalists2007/bingham.cfm (accessed August 28, 2009).
53. www.interfaithpowerandlight.org (accessed August 28, 2009).
54. Virginian Interfaith Power and Light brochure, distributed at Bon Air Presbyterian Church, Richmond, Virginia, at its Earth Day observance, April 29, 2007.
55. King, p. 54.
56. Max Oelschlaeger, *Caring for Creation: An Ecumenical Approach to the Environmental Crisis* (Geneva: World Council of Churches, 1994), p. 222. Quoted in King, pp. 54–55.
57. Jo Freeman, "The Women's Liberation Movement: Its Origin, Structures and Ideals" (Pittsburgh, PA: Know, 1971). http://scriptorium.lib.duke.edu/wlm/womlib (accessed August 28, 2009).
58. John Jefferson Davis, "Ecological 'Blind Spots' in the Structure and Content of Recent Evangelical Systematic Theologies," *Journal of the Evangelical Theological Society* 43/2 (June 2000), p. 273.
59. Ibid., p. 275.
60. "Moyers on America: Is God Green?" Films for the Humanities and Sciences (Princeton, NJ: Films Media Group, n.d.).
61. Leslie Scanlon, "People of faith concerned about ecology; churches going 'green,'" *The Presbyterian Outlook*, Feb. 26, 2007, pp. 8–9.
62. Ibid.
63. Ibid.
64. www.christiansandclimate.org/statement (accessed August 28, 2009).
65. King, p. 40.
66. The term "creation care" signals the approach of evangelical Christians to the relationship between Christians and the earth. It is also

associated with the terms "environmental stewardship" and "earth keeping."
67. www.ausable.org/au.ausableidea.cfm (accessed August 28, 2009).
68. www.creationcsp.org/about.html (accessed August 28, 2009).
69. www.creationcare.org/magazine/winter07.php (accessed August 28, 2009).
70. Other initiatives include the Ecumenical Water Network, http://water.oikoumene.org and a variety of programs of the World Council of Churches.
71. www.pastornet.net.au/envcomm/News/Anglican%20News%20Columns/ANMAR07.pdf (accessed August 28, 2009).
72. www.acen.anglicancommunion.org/reports/2005.cfm.
73. (Cambridge, MA: Harvard University Press, 2007).
74. Sarah McFarland Taylor, p. 5.
75. www.genesisfarm.org (accessed August 28, 2009).
76. Roger Gottlieb, *A Greener Faith: Religious Environmentalism and Our Planet's Future* (New York: Oxford University Press, 2006), pp. 186–187.
77. www.columbiariver.org (accessed August 28, 2009).
78. Ibid.
79. See also, Andrew Nikiforuk, *Tar Sands: Dirty Oil and the Future of a Continent* (Vancouver: Greystone Books, 2009); and "Like Oil and Water: the true cost of the tar sands 2," www.kairoscanada.org/e/action/ LikeOil&Water_AlbertaStory.pdf (accessed August 28, 2009).
80. "The Integrity of Creation and the Athabasca Oil Sands." Available at www.dioceseofstpaul.ca/index.php?option=com_content7task=view7id=135 (accessed August 28, 2009).

BIBLIOGRAPHY

Allen, Paula Gunn. *Grandmothers of the Light: A Medicine Woman's Source Book.* Boston: Beacon Press, 1991.

Allenby, Brad. "The Anthropogenic Earth: Integrating and Reifying Technology, Environmentalism, and Religion," *Princeton Seminary Bulletin* 24, no. 1 (2003): 104–121.

Anderson, Alan C., and Philip N. Joranson. *Religious Reconstruction for the Environmental Future: An Interdisciplinary Workshop to Confront the Ethical Urgency at the Bishop Center for Christian Education of the University of Connecticut at Storrs, Conn.* South Coventry: FMN Group (1973).

Anderson, Benedict. *Imagined Communities: Reflections on the Origin and Spread of Nationalism.* Rev. ed. London: Verso, 1983.

Anderson, Gary. "Stewardship of the Earth: A Christian Response to a Spiritual Crisis." *Epiphany* 3, no. 3 (1983): 2–109.

Appenzeller, Tim. "The End of Cheap Oil." *National Geographic* (June 2004): 80–109.

Appleby, R. Scott, and Martin E. Marty, eds. *Fundamentalisms and Society: Reclaiming the Sciences, the Family, and Education.* Chicago: University of Chicago Press, 1993.

Aquino, Maria Pilar. *Our Cry for Life: Feminist Theology from Latin America.* Maryknoll, NY: Orbis, 1994.

Association of Baltimore Area Grantmakers. "Baby Boomers Give More Than Older Americans." *The Chronicle of Philanthropy* (2005).

Bainbridge, William. "Computer Simulation of Cultural Drift." *Journal of the British Interplanetary Society* 37 (1984): 420–429.

Bakken, Peter, and Steven Bouma-Prediger, eds. *Evocations of Grace: Writings of Joseph Sittler on Ecology, Theology, and Ethics.* Grand Rapids, MI: Eerdmans, 2000.

Balabanski, Vicky, and Norman C. Habel, eds. *The Earth Story in the New Testament*. Vol. 5. The Earth Bible. London: Sheffield Academic Press, 2002.

Barbour, Ian G., ed. *Earth Might Be Fair: Reflections on Ethics, Religion and Ecology*. Englewood Cliffs, NJ: Prentice-Hall, 1972.

Barlow, Connie. *Green Space, Green Time: The Way of Science*. New York: Copernicus, 1997.

Barnette, Henlee H. *The Church and the Ecological Crisis*. Grand Rapids, MI: Eerdmans, 1972.

Barthes, Roland. *Mythologies*, trans. Annette Lavers. New York: Hill and Wang, 1972.

Bayes, Jane H., and Nayereh Tohidi, eds. *Globalization, Gender and Religion: The Politics of Women's Rights in Catholic and Muslim Contexts*. Waterloo, ON: Wilfrid Laurier University Press, 2001.

Beaumont, Tim. "Options for the 90s: From Environmentalism to Ecology." *Modern Churchman* 28, no. 1 (1985): 11–16.

Bennett, John B. "On Responding to Lynn White: Ecology and Christianity." *Ohio Journal of Religious Studies* 5 (1977): 71–77.

Bergant, Dianne. *The Earth Is the Lord's: The Bible, Ecology, and Worship*. American Essays in Liturgy, ed. Edward Foley. Collegeville, MN: Liturgical Press, 1998.

Bergmann, Sigurd. *Creation Set Fire: The Spirit as Liberator of Nature*, trans. Douglas Stott. Grand Rapids, MI: Eerdmans, 2005.

Berleant, Allen, and Arnold Carlson, eds. *The Aesthetics of Natural Environment*. New York: Broadview Press, 2004.

Berry, Thomas. "The North American Continent." An unpublished poem.

———. "Perspectives on Creativity: Openness to a Free Future." In *Whither Creativity, Freedom, Suffering*, 1–24. Villanova, PA: Villanova University Press, 1981

———. *The Dream of the Earth*. San Francisco: Sierra Club Books, 1988.

———. *The Great Work: Our Way into the Future*. New York: Bell Tower, Random House, 1999.

———. "Every Being Has Rights." In *Twenty-Third Annual E.F. Schumacher Lectures*, ed. Hildegrade Hannum, 29. Stockbridge, MA: E.F. Schumacher Society, 2003.

———. "Into the Future." In *This Sacred Earth: Religion Nature and Environment*, ed. Roger S. Gottlieb, 494–496. New York: Routledge, 2004.

———. "Contemporary Spirituality: The Journey of the Human Community." *Cross Currents* (1974): 172–183.

Berry, Wendell. *The Gift of Good Land: Further Essays Cultural and Agricultural*. San Francisco: North Point, 1981.
Berryman, Philip. *Liberation Theology: Essential Facts about the Revolutionary Movement in Latin America and Beyond*. New York: Pantheon, 1986.
Birch, Charles. "Creation, Technology and Human Survival: Called to Replenish the Earth." *Ecumenical Review* 28 (1976): 66–79.
———. "Nature, God and Humanity in Ecological Perspective." *Christianity and Crisis* 39, no. 29 (1979): 259–266.
Birch, Charles, William R. Eakin, and Jay B. McDaniel, eds. *Liberating Life: Contemporary Approaches to Ecological Theology*. Maryknoll, NY: Orbis, 1990.
Birch, Charles, and John B. Cobb Jr. *The Liberation of Life: From the Cell to the Community*. Cambridge, MA: Cambridge University Press, 1981.
Blurton Jones, Nicholas G., Kristen Hawkes, and James F. O'Connell. "Antiquity of Postreproductive Life: Are There Modern Impacts on Hunter-Gatherer Postreproductive Life Spans?" *American Journal of Human Biology* 14 (2002): 185.
Bocking, Stephen, ed. *Biodiversity in Canada: Ecology, Ideas, and Action*. Peterborough: Broadview Press, 2000.
Boff, Leonardo. *Ecology and Theology: A New Paradigm*. Maryknoll, NY: Orbis, 1995.
———. *Cry of the Earth, Cry of the Poor*. Maryknoll, NY: Orbis, 1997.
Bonifazi, Conrad. *A Theology of Things: A Study of Man in His Physical Environment*. Philadelphia: Lippincott, 1967.
Bookless, David. "Between the Rock and a Hard Place: The Developing Work of a Rocha." *Ecotheology* 7, no. 2 (2003): 213–220.
Bouchard, Bishop Luc. "The Integrity of Creation and the Athabasca Oil Sands." Pastoral Letter. January 25, 2009. Available at www.dioceseofstpaul.ca/ index.php?option=com_content&task=view&id=135 (accessed August 28, 2009).
Boulding, Kenneth Ewart. "Science and the Christian Phylum in Evolutionary Tension." In *Experiment of Life*, 89–109. Toronto: University of Toronto Press, 1983.
Bouma-Prediger, Steven. *The Greening of Theology: The Ecological Models of Rosemary Radford Ruether, Joseph Sittler, and Jurgen Moltmann*. Atlanta: Scholars Press, 1995.
———. *For the Beauty of the Earth: A Christian Vision for Creation Care*. Grand Rapids, MI: Baker Academic, 2001.

Bowers, C. A. *The Culture of Denial: Why the Environmental Movement Needs a Strategy for Reforming Universities and Public Schools*. Albany, NY: State University of New York Press, 1997.
Boyle, Joseph. "Natural Law," in *The New Dictionary of Theology*. Eds. Joseph A. Komonchak, Mary Collins, and Dermot A. Lane. Wilmington, DE: Michael Glazier, 1988, 703–708.
Bradley, Ian C. *God Is Green: Christianity and the Environment*. London: Darton, Longman & Todd, 1990.
Bragg, Wayne. "Beyond Development." In *Church in Response to Human Need*, 37–95. Monrovia: Missions Advanced Research & Communication Center, 1983.
Bramwell, Anna. *Ecology in the 20th Century: A History*. New Haven: Yale University Press, 1989.
Brooks, Paul. *The House of Life: Rachel Carson at Work*. Boston: Houghton Mifflin, 1972.
Brown, Harrison. *The Human Future Revisited: The World Predicament and Possible Solutions*. New York: W.W. Norton, 1978.
Brown, Lester. *The Twenty-Ninth Day: Accommodating Human Needs and Numbers to the Earth's Resources*. New York: W.W. Norton, 1978.
Brunk, Conrad, and Harold Coward, eds. *Acceptable Genes? Religious Traditions and Genetically Modified Foods*. Albany, NY: State University of New York, 2009.
Bryce-Smith, D. "Ecology, Theology and Humanism." *Zygon* 12 (1977): 212–231.
Bullock, Wilbur L. "Brother Earth: Nature, God and Ecology in Time of Crisis." *Christianity Today* 15 (1971): 20–24.
———. "Ecology and Religion: Toward a New Christian Theology of Nature." *Christian Scholars' Review* 14, no. 3 (1985): 279–280.
Cameron, Anne. "First Mother and the Rainbow Children." In *Healing the Wounds: The Promise of Ecofeminism*, ed. Judith Plant. Philadelphia: New Society Press, 1993.
Caputi, Jane. "Nuclear Power and the Sacred or Why a Beautiful Woman Is Like a Nuclear Plant." In *Ecofeminism and the Sacred*, ed. Carol J. Adams, 229–250. New York Continuum, 1993.
Carson, Rachel. *Silent Spring*. Boston: Houghton Mifflin, 1962.
Castoriadis, Cornelius. *The Imaginary Institution of Society*. Cambridge, UK: Polity Press, 1997.
Census FactFinder. http://factfinder.census.gov/servlet/ThematicMap FramesetServlet? bm=y&-geo_id=01000US&tm_name=ACS_ 2005_EST_G00_M00645&-ds_name=ACS_2005_

EST_G00_&-_MapEvent=displayBy&-_dBy=040 (accessed August 27, 2009).
Charney, J. G. *Carbon Dioxide and Climate: A Scientific Assessment.* Washington, DC: National Academy of Sciences, 1979.
Chaves, Mark. *Congregations in America.* Cambridge, MA: Harvard University Press, 2004.
Chopp, Rebecca S. *The Praxis of Suffering: An Interpretation of Liberation and Political Theologies.* Maryknoll, NY: Orbis, 1986.
Christ, Carol P. *She Who Changes: Re-Imagining the Divine in the World.* New York: Palgrave Macmillan, 2003.
Christiansen, Drew, and Walter Grazer, eds. *And God Saw That It Was Good: Catholic Theology and the Environment.* Washington, DC: United States Catholic Conference, 1996.
Chryssavgis, John. *Cosmic Grace, Humble Prayer: The Ecological Vision of the Green Patriarch Bartholomew I.* Grand Rapids, MI: Eerdmans, 2003.
Cladwell, Lynton. *International Environmental Policy: From the Twentieth to the Twenty First Century.* 3rd ed. Durham, NC: Duke University Press, 1996.
Clayton, Susan, and Susan Opotow, eds. *Identity and the Natural Environment: The Psychological Significance of Nature.* Cambridge, MA: MIT Press, 2003.
Cobb, John B. Jr. "A Critical View of Inherited Theology." *Christian Century* (1980): 194–197.
———. *The Greening of Theology,* www.religion-online.org/ showarticle.asp?title=1093 (accessed June 14 2006).
———. *Is It Too Late? A Theology of Ecology.* Rev. ed. Denton: Environmental Ethics Books, 1995.
———. "The Making of an Earthist Christian." In *Encyclopedia of Religion and Nature,* 1.
———. *Sustainability: Economics, Ecology and Justice.* Eugene, OR: Wipf and Stock, 2007.
———. "The Road to Sustainability: Progress and Regress." http://religion-online.org/showarticle.asp?title=3375 (accessed August 28, 2009).
Commoner, Barry. *The Closing Circle.* New York: Knopf, 1971.
Conradie, Ernst M. *An Ecological Christian Anthropology: At Home on Earth?* Hants, UK: Ashgate, 2005.
Conradie, Ernst M., and Julia Martin. "Gender, Religion and the Environment: A University of the Western Cape Case Study." *Ecotheology* 11, no. 4 (2006): 431–444.

Cooper, Mary Anderson. "Welfare's Role in Creating a Society for All Ages: Recognizing the Multigenerational Impact of Welfare Reform." *Church & Society* (1999).

Cooper, Tim. *Green Christianity*. London: Hodder & Stoughton, 1990.

Cosgrove, Denis. "Contested Global Visions: One-World, Whole-Earth, and the Apollo Space Photographs." *Annals of the Association of American Geographers* 84, no. 2 (1994): 270–294.

Coward, Harold, Rosemary Ommer, and Tony Pitcher, eds. *Just Fish: Ethics and Canadian Marine Fisheries*. St John's: Centre for Studies in Religion and Society, 2000.

Coward, Harold, and Daniel C. Maguire, eds. *Visions of a New Earth: Religious Perspectives on Population, Consumption, and Ecology*. Albany, NY: State University of New York Press, 2000.

Coward, Harold, and Andrew J. Weaver, eds. *Hard Choices: Climate Change in Canada*. Waterloo, ON: Wilfred Laurier University Press, 2004.

Cox, Louis. *Truth Is Something That Happens*. www.quakerearthcare.org/ SpiritualityandEarthcare/index.htm (accessed August 27, 2009).

Cuénot, Claude. *Teilhard de Chardin*. Baltimore, MD: Helicon Press, 1965.

Dale, Ann. *At the Edge: Sustainable Development in the 21st Century*. Vancouver, BC: University of British Columbia Press, 2001.

Dalton, Anne Marie. "Who Cares About the Meadow?" In *Every Grain of Sand: Canadian Perspectives on Ecology and the Environment*, ed. Andrew Wainwright, 73–85. Waterloo, ON: Wilfrid Laurier University Press, 2004.

———. *A Theology for the Earth: The Contributions of Thomas Berry and Bernard Lonergan*. Ottawa: University of Ottawa Press, 1999.

———. "Gender and Ecofeminism: Religious Reflections on a Case Study in Soc Son, Vietnam." *Ecotheology* 11, no. 4 (2006): 398–414.

Daly, Herman E., and John B. Cobb Jr. *For the Common Good: Redirecting the Economy toward Community, the Environment and a Sustainable Future*. 2nd ed. Boston: Beacon Press, 1994.

Darling-Smith, Barbara. "Rosemary Radford Ruether." In *Encyclopedia of Religion and Nature*, ed. Bron Taylor, 2, 1433–1434. New York: Thoemmes Continuum, 2005.

Davis, John Jefferson. "Ecological 'Blind Spots' in the Structure and Content of Recent Evangelical Systematic Theologies." *Journal of the Evangelical Theological Society* 43, no. 2 (2000): 273–278.

Deane-Drummond, Celia. *A Handbook in Theology and Ecology.* London: SCM Press, 1996.
———. *Gaia and Green Ethics: Implications of Ecological Theology.* Cambridge, UK: Grove Books, 2000.
———. *Genetics and Christian Ethics.* Cambridge, UK: Cambridge University Press, 2006.
Deloria, Vine. "God Is Also Red: An Interview with Vine Deloria Jr. by James Mcgraw." *Christianity and Crisis* 35, no. 15 (1975): 198–206.
DeMarinis, Valerie, and Michael B. Aune, eds. *Religious and Social Ritual: Interdisciplinary Explorations.* Albany, NY: State University of New York Press, 1996.
Derr, Thomas Sieger. *Ecology and Human Need.* Philadelphia: Westminster Press, 1973.
———. "Religion's Responsibility for the Ecological Crisis: An Argument Run Amok." *Worldview* 18 (1975): 39–45.
Derrick, Christopher. *The Delicate Creation: Towards a Theology of the Environment.* Old Greenwich: Devin-Adair, 1972.
DeSteiguer, Joseph Edward. *The Origins of Modern Environmental Thought.* Tucson: University of Arizona Press, 2006.
DeWitt, Calvin. "Planetary Justice: Christians Interested in Environmental, Interview by Randy Frame." *Christianity Today* 32 (1988): 74.
———. "The Scientist and the Shepherd." In *The Oxford Handbook of Religion and Ecology*, ed. Roger Gottlieb. New York: Oxford University Press, 2006.
Diamond, Irene. "Toward a Cosmology of Continual Creation: From Ecofeminism to Feminine Ecology and Umbilical Ties." *Cross Currents* 54, no. 4 (2004): 7–16.
Diamond, Irene, and Gloria Feman Orenstein, ed. *Reweaving the World: The Emergence of Ecofeminism.* San Francisco: Sierra Club Books, 1990.
Di Chiro, Giovanna. "Nature as Community: The Convergence of Environment and Social Justice." In *Privatizing Nature: Political Struggles for the Global Commons*, ed. Michael Goldman. New Brunswick, NJ: Rutgers University Press, 1998.
Donaldson, James. "America the Beautiful: Interdependence with Nature." *Foundations* 19 (1976): 238–256.
Dunn, Stephen. "Three Shades of Green, a Canadian Experiment." Paper presented at Ecotheology Conference. Chicago, 1995.
Earth Charter in Action, www.earthcharterinaction.org/assets/pdf/charter/charter_eng.pdf (accessed August 27, 2009).

Eaton, Heather. *Introducing Ecofeminist Theologies*. London: T&T Clark International, 2005.
Eaton, Heather, and Lois Lorentzen, eds. *Ecofeminism and Globalization*. Lanham, MD: Rowman & Littlefield, 2003.
Edwards, Denis. *Ecology at the Heart of Faith: The Change of Heart That Leads to a New Way of Living on Earth*. Maryknoll, NY: Orbis, 2007.
Eliade, Mircea. *The Sacred and the Profane: The Nature of Religion*. 2nd ed. Trans. Willard R. Trask. New York and London: Harcourt Brace Jovanovich, 1959.
Ellison, Christopher, and Darren Sherkat. "Structuring the Religion-Environment Connection: Identifying Religious Influences on Environmental Concern and Activism." *Journal for the Scientific Study of Religion* 46, no. 1 (2007): 71–86
Erhard, Nancie. *Moral Habitat. Ethos and Agency for the Sake of the Earth*. Albany, NY: State University of New York Press, 2007.
Errington, Frederick K., and Deborah Gewertz. "The Chief of the Chambri: Social Change and Cultural Permeability among a New Guinea People." *American Ethnologist* 12, no. 3 (1985): 442–454.
Exploring Sustainable Development: Global Scenarios 2000–2005. Geneva: World Business Council for Sustainable Development, 1997.
Faithlink. www.faithlink.com (accessed August 28, 2009).
Falk, Richard. *This Endangered Planet*. New York: Random House, 1971.
Ferkiss, Victor. "Christianity, Technology and the Human Future." *Dialog: A Journal of Theology* 13 (1974): 258–263.
———. "Nature, Technology, and Politics in a Global Context." *Zygon* 16 (1981): 127–149.
Ferre, Frederick Pond. "Religious World Modeling and Postmodern Science." *Journal of Religion* 62, no. 3 (1982): 261–271.
———. "God and Global Community." In *God and Global Justice*, 3–16. New York: Paragon House, 1985.
Finger, Thomas N. *Self, Earth and Society: Alienation and Trinitarian Transformation*. Downers Grove, IL: InterVarsity Press, 1997.
Fowler, Robert Booth. *The Greening of Protestant Thought*. Chapel Hill: University of North Carolina Press, 1995.
Frame, Randall L. "Christianity and Ecology: A Better Mix Than Before." *Christianity Today* 34 (1990): 38–39.
Freeman, Martha, ed. *Always, Rachel: The Letters of Rachel Carson and Dorothy Freeman 1952–1964*. Boston: Beacon Press, 1995.

Frickel, Scott, and Neil Gross. "A General Theory of Scientific Intellectual Movements." *American Sociological Review* 70 (2005): 204–232.

Friends Committee on National Legislation. "Over 160 Congregations in the Greater Washington Area Engage in Efforts to Reduce the Dangers of Global Warming." www.fcnl.org/press/releases/globalwarm092606.htm (accessed August 25, 2009).

Gaard, Greta, ed. *Ecofeminism: Women, Animals, Nature*. Philadelphia: Temple University Press, 1993.

Gaffield, Chad, and Pam Gaffield, eds. *Consuming Canada: Readings in Environmental History*. Toronto: Copp Clark, 1995.

Gebara, Ivone. *Longing for Running Water: Ecofeminism and Liberation*. Minneapolis: Fortress Press, 1999.

———. "Ecofeminism: A Latin American Perspective." *Cross Currents* 53, no. 1 (2003): 93–105

Genesis Farm. www.genesisfarm.org (accessed August 28, 2009).

Gibson, William E. "Ecojustice: Burning Word: Heilbroner and Jeremiah to the Church." *Foundations* 20, no. 5 (1977): 318–328.

———. "Praise the Creator—Care for Creation." In *Social Themes of the Christian Year*, 236–242. Philadelphia: Geneva Press, 1983.

———. "Beginning a 'Turnaround Decade': The Many Faces of Earth Day." *Christianity and Crisis* 50 (1990): 147–149.

Gillis, Peter, and Thomas R. Roach, "The Beginnings of a Movement: The Montreal Congress and its Aftermath, 1880–1896." In Gaffield and Gaffield, eds. *Consuming Canada*, 131–151.

Global Future: Time to Act: Report to the President on Global Resources, Environment, and Population. Washington, DC: U.S. Council on Environmental Quality, 1981.

Gnanadason, Aruna. *Listen to the Women! Listen to the Earth!* Geneva: WCC Publications, 2005.

———. "Yes, Creator God, Transform the Earth! The Earth as God's Body in an Age of Environmental Violence." *Union Seminary Quarterly Review* 58, no. 1–2 (2005): 107–124.

Goldman, Michael, ed. *Privatizing Nature: Political Struggles for the Global Commons*. New Brunswick, NJ: Rutgers University Press, 1998.

Gosling, David L. "Scientific Decisions and Social Goals." In *Man in Nature*, 45–58. Colombo, Sri Lanka: Ecumenical Institute for Study and Dialogue, 1979.

———. "Towards a Credible Ecumenical Theology of Nature." *Ecumenical Review* 38, no. 3 (1986): 322–331.

Gottlieb, Roger S., ed. *This Sacred Earth: Religion, Nature, Environment*. New York: Routledge, 2004.

———. *A Greener Faith. Religious Environmentalism and Our Planet's Future.* New York: Oxford University Press, 2006.
"Grandmother Hypothesis." www.aka-alias.net/2004/09/grandmother-hypothesis.html (accessed August 28, 2009).
Gratton, Brian, and Carole Haber. "Three Phases in the History of American Grandparents: Authority, Burden, Companion," *Generations* 20, no. 1 (Spring, 1996): 7–12.
Gray, Elizabeth Dodson. *Adam's World*. Video. National Film Board of Canada, 1989.
Green, Lorna. *Earth Age: A New Vision of God, the Human and the Earth*. New York: Paulist Press, 2003.
Gregersen, Niels Henrik. "Grace in Nature and History: Luther's Doctrine of Creation Revisited." *Dialog: A Journal of Theology* 44, no. 1 (2005): 19–29.
———. "The Complexification of Nature: Supplementing the Neo-Darwinian Paradigm?" *Theology and Science* 4, no. 1 (2006): 5–31.
Greisch, Janet Rohler. "Ecology Crisis: God's Creation and Man's Pollution." *Christianity Today* 15 (1971).
Grey, Mary, and Elizabeth Green, eds. *Ecofeminism and Theology*. Vol. 2, Yearbook of the European Society of Women in Theological Research. Kampen, the Netherlands: Kok Pharos, 1994.
Groshong, Kimm. "The Noisy Response to Silent Spring; Placing Rachel Carson's Work in Context." Senior Thesis, 2008. Claremont, CA: Pomona College, 2002. www.sts.pomona.edu/Thesis-STS.pdf (accessed November 23, 2009).
Gross, Rita M., and Rosemary Radford Ruether. *Religious Feminism and the Future of the Planet: A Christian-Buddhist Conversation*. New York: Continuum International Publishing Group, 2001.
Guardini, Romano. www.jknirp.com/guard18.htm (accessed August 28, 2009).
Guha, Ramachandra. *Environmentalism: A Global History*, Longman World History Series. Ed. Michael Adas. New York: Addison Wesley Longman, 2000.
Habel, Norman C. *Readings from the Perspective of Earth*. The Earth Bible, vol. 1, ed. Norman C. Habel. Cleveland, OH: Pilgrim Press, 2001.
———. *The Earth Story in the Psalms and the Prophets*. The Earth Bible, vol. 4. Cleveland, OH: Pilgrim, 2001.
Habel, Norman C., and Shirley Wurst, ed. *The Earth Story in Genesis*. The Earth Bible, vol. 2. Sheffield: Sheffield Academic Press, 2000.

Hall, Douglas John. *Professing the Faith: Christian Theology in a North American Context.* Minneapolis: Fortress Press, 1996.

Hallman, David C. *A Place in Creation: Ecological Visions in Science, Religion and Economics.* Toronto: United Church of Canada Publishing House, 1992.

———, ed. *Ecotheology: Voices from the South and North.* Geneva: WCC Publications, 1994.

Harakas, Stanley S. "The Integrity of Creation and Ethics." *St. Vladimir's Theological Quarterly* 32, no. 1 (1988): 27–42.

Hardin, Garrett. *Exploring New Ethics for Survival: The Voyage of the Spaceship Beagle.* New York: Viking Press, 1972.

———. "Ecology and the Death of Providence." *Zygon* 15, no. 1 (1980): 57–68.

Hargrove, Eugene C., ed. *Religion and the Environmental Crisis.* Athens, GA: University of Georgia Press, 1986.

Harvey, Graham. "Paganism—Contemporary." In *Encyclopedia of Religion and Nature*, ed. Bron R. Taylor. London: Thoemmes Continuum, 2005.

Haught, John. *The Cosmic Adventure: Science, Religion, & the Quest for Purpose.* New York/Ramsey, NJ: Paulist Press, 1984.

———. *Deeper Than Darwin: The Prospect for Religion in the Age of Evolution.* Boulder, CO: Westview Press, 2003.

———. *Is Nature Enough? Meaning and Truth in the Age of Science.* Cambridge, UK: Cambridge University Press, 2006.

———. *The Promise of Nature: Ecology and Cosmic Purpose.* New York: Paulist Press, 1993.

Hayes, Bernadette, and Manussos Marangudakis. "Religion and Environmental Issues within Anglo-American Democracies." *Review of Religious Research* 42, no. 2 (2000): 159–174.

Hays, Samuel P. "From Conservation to Environment: Environmental Politics in the United States Since World War II." *Environmental Review* 6, no. 2 (Fall, 1982): 14–29.

Heimann, P. M. "Voluntarism and Immanence; Conceptions of Nature in Eighteenth-Century Thought." In *Philosophy, Religion and Science in the 17th and 18th Centuries, Library of the History of Ideas*, vol. 2, ed. John W. Yolton. Rochester, NY and Woodbridge, Suffolk: University of Rochester Press, 1990.

Heller, Chaia. *Ecology of Everyday Life: Rethinking the Desire for Nature.* Montreal, New York, London: Black Rose Books, 1999.

Henry, Granville C. "The Image of a Machine in the Liberation of Life." *Process Studies* 13 (1983): 143–153.

Henry, Granville C., and Dean R. Fowler. "World Models." *Journal of the American Academy of Religion* 42 (1974): 114–127.

Hersch, Joni, and W. Kip Viscusi. "The Generational Divide in Support for Climate Change Policies: European Evident." In *Harvard Law and Economics Discussion Paper No. 504*, February 2005.

Hervieu-Léger, Daniele. "Apocalyptique Ecologique et 'Retour' de la Religion." *Archives de Science Sociales des Religion* 27 (1982): 49–67.

Hessel, Dieter T., and Rosemary Radford Ruether, eds. *Christianity and Ecology*. Cambridge, MA: Harvard University Press, 2000.

Hessel, Dieter T., ed. *Theology for Earth Community: A Field Guide*. Maryknoll, NY: Orbis, 1996.

———. "Where Were/Are the U.S. Churches in the Environmental Movement?" In *Theology for the Earth Community: A Field Guide*, ed. Dieter T. Hessel. Maryknoll, NY: Orbis, 1996.

Hessel, Dieter, and Larry Rasmussen, eds. *Earth Habit: Eco-Justice and the Church's Response*. Minneapolis: Augsburg Fortress, 2001.

Hillel, Daniel. *The Natural History of the Bible: An Environmental Exploration of the Hebrew Scriptures*. New York: Columbia University Press, 2006.

Holdgate, Martin W., Mohammed Kassas, and Gilbert F. White. *The World Environment 1972–1982*. Nairobi: United Nations Environment Programme, 1982.

hooks, bell. *Ain't I a Woman? Black Women and Feminism*. Boston: South End Press, 1981.

Houston, James M. "Environmental Movement: Five Causes of Confusion." *Christianity Today* 16 (1972): 8–10.

IHM Sisters. www.ihmsisters.org/www/Sustainable_Community/Sustainable_Renovation/sustainmep.asp (accessed August 28, 2009).

Irigaray, Luce. *Democracy Begins between Two*, trans. Kirsteen Anderson. New York: Routledge, 2001.

Isasi-Díaz, Ada María. *Mujerista Theology*. Maryknoll, NY: Orbis, 1996.

IUCN. *World Conservation Strategy: Living Resource Conservation for Sustainable Development*. Gland, Switzerland: International Union for Conservation of Nature and Natural Resources, 1980.

Ivakhiv, Adrian. "Religion, Nature and Culture: Theorizing the Field." *Journal for the Study of Religion, Nature and Culture* 1, no. 1 (2007): 47–57.

Ives, Christopher, and John B. Cobb Jr., eds. *The Emptying God: A Buddhist-Jewish-Christian Conversation*. Eugene, OR: Wipf and Stock, 1990.

Jacoby, Karl. "Conservation." In *Encyclopedia of World Environmental History*, vol. 1. Eds. Shepard Krech III, J. R. McNeil, and Carolyn Merchant. New York and London: Routledge, 2004, 262–268.

Jasen, Patricia. *Wild Things: Nature, Culture and Tourism in Ontario, 1790–1914*. Toronto: University of Toronto Press, 1995.

Jensen, Derrick. *Listening to the Land: Conversations about Nature, Culture and Eros*. San Francisco: Sierra Club Books, 1995.

Johnson, Elizabeth. *She Who Is: The Mystery of God in Feminist Theological Discourse*. New York: Crossroad, 1993.

Johnson, Patricia-Anne. "Womanist Theology as Counter-Narrative." In *Gender, Ethnicity and Religion: Views from the Other Side*, ed. Rosemary Radford Ruether. Minneapolis: Fortress Press, 2002, 197–214.

Johnson, Theodore W. "Current Thinking on Size Transition." *Size Transitions in Congregations*, ed. Beth Ann Gaede, 3–30. Washington, DC: Alban Institute, 2001.

Johnston, Carol. *The Wealth or Health of Nations: Transforming Capitalism from Within*. Cleveland: Pilgrim Press, 1998.

Jung, Lee Hong. "Healing and Reconciliation as the Basis for the Sustainability of Life: An Ecological Plea for a 'Deep' Healing and Reconciliation." *International Review of Mission* 94, no. 372 (2005): 85–102.

Jung, Shannon. *We Are Home: A Spirituality of the Environment*. Mahwah, NJ: Paulist Press, 1993.

KAIROS: Canadian Ecumenical Justice Initiatives www.united-church.ca/Partners/ecumenical/kairos (accessed August 31, 2009):

Keller, Catherine. "Eschatology, Ecology, and a Green Ecumenacy." *Ecotheology* 2 (1997): 84–99.

———. "The Lost Fragrance: Protestantism and the Nature of What Matters." *Journal of the American Academy of Religion* 65 (1997): 355–370.

———. *Face of the Deep: A Theology of Becoming*. New York: Routledge, 2003.

———. *God and Power: Counter Apocalyptic Journeys*. Minneapolis: Augsburg, 2005.

Keller, Catherine, and Laurel Kearns, eds. *Ecospirit: Religions and Philosophies for the Earth*. New York: Fordham University Press, 2007

Kerans, Pat, and John Kearney. *Turning the World Right-Side Up: Science, Community and Democracy*. Halifax, NS: Fernwood, 2006.

Keulartz, Jozef. *The Struggle for Nature: A Critique of Radical Ecology*, trans. Rob Kuitenbrouwer. London: Routledge, 1998.

King, Carolyn M. "Ecotheology: A Marriage between Secular Ecological Science and Rational, Compassionate Faith." *Ecotheology* 10 (2001): 40–69.

———. *Habitat of Grace: Biology, Christianity and the Global Environmental Crisis*. ATF Science and Theology Series, vol. 3, ed. Mark Wm. Worthing. Adelaide: Openbook, 2002.

Kornhaber, Arthur. *Contemporary Grandparenting*. Thousand Oaks, CA: Sage, 1996.

Krech, Shepard III. *The Ecological Indian: Myth and History*. New York: W.W. Norton, 1999.

Kroll-Smith, J. Stephan, and Stephan Robert Couch. "A Chronic Technical Disaster and the Irrelevance of Religious Meaning: The Case of Centralia Pennsylvania." *Journal for the Scientific Study of Religion* 26, no. 1 (1987): 25–37.

Kwok Pui-lan. *Introducing Asian Feminist Theology*. Introductions in Feminist Theology, eds. Mary Grey, et al. Sheffield, UK: Sheffield Academic Press, 2000.

Kyung, Chung Hyun. "Ecology, Feminism and African and Asian Spirituality: Toward a Spirituality of Eco-Feminism." In *Ecotheology: Voices from South and North*, ed. David G. Hallman. Maryknoll, NY: Orbis, 1994, 175–178.

———. *Struggle to Be Sun Again: Introducing Asian Women's Theology*. Maryknoll, NY: Orbis, 1994.

LaBar, Martin. "Message to Polluters from the Bible." *Christianity Today* 18 (1974): 8–12.

Lahdenperä M., V. Lummaa, and A. F. Russell. "Menopause: Why Does Fertility End before Life?" *Climacteric* 7, no. 4 (2004): 327–331.

Lampman, Jane. "Churches Go Green." In *Christian Science Monitor*, October 23, 2003, www.csmonitor.com/2003/0123/p11s02-lire.html (accessed August 28, 2009).

Lane, Belden. *The Solace of Fierce Landscapes: Exploring Desert and Mountain Spirituality*. New York: Oxford University Press, 1998.

Latour, Bruno. *Politics of Nature: How to Bring the Sciences into Democracy*, trans. Catherine Porter. Cambridge, MA: Harvard University Press, 2004.

Lawton, Kim A. "Is There Room for Profile Environmentalists." *Christianity Today* 34 (1990): 46–47.

Leiss, William. *The Domination of Nature*. New York: Braziller, 1972.

Lester, Rita. "The Nature of Nature: Ecofeminism and Environmental Racism in America." *Gender, Ethnicity & Religion*, ed. Rosemary Radford Ruether. Minneapolis: Fortress Press, 2002, 230–246.

Levin, Simon. *Fragile Dominion: Complexity and the Commons*. Reading, MA: Helix Books, Perseus Books, 1999.

Lonergan, Anne, and Caroline Richards, eds. *Thomas Berry and the New Cosmology*. Mystic, CT: Twenty-Third Publications, 1987.

Lonergan, Bernard. *Insight: A Study in Human Understanding*. San Francisco: Harper & Row, 1978.

———. *Method in Theology*. New York: Herder & Herder, 1972.

Lorde, Audre. *Zami: A New Spelling of My Name*. Freedom, CA: Crossing Press, 1982.

Low, Alaine, and Soraya Tremyane, eds. *Women as Sacred Custodians of the Earth? Women, Spirituality and the Environment*. New York: Berghahn Books, 2001.

Low, Mary. *Celtic Christianity and Nature: Early Irish and Hebridean Traditions*. Edinburgh: Edinburgh University Press, 1996.

Lynch, Timothy B. "Two Worlds Join to 'Preserve the Earth'." *Christianity and Crisis* 50 (1990): 142–144.

MacGregor, Sherilyn. *Beyond Mothering Earth: Ecological Citizenship and the Politics of Care*. Vancouver: University of British Columbia Press, 2006.

Manolopoulis, Mark. "Derrida's Gift to Eco/Theo/Logy: A Critical Tribute." *Cross Currents* 54, no. 4 (2005): 55–68.

Man's Impact on the Global Environment: Assessment and Recommendations for Action. Cambridge, MA: Study of Critical Environmental Problems, 1970.

Marshall, Paul. "A Christian View of Economics." *Crux* 21, no. 1 (1985): 3–6.

McDaniel, Jay, "Christianity (7f)-Process Theology." In *Encyclopedia of Religion and Nature*, vol. 1, 364–366.

———. "Christianity and the Need for a New Vision." In *Religion and the Environmental Crisis*, 188–212. Athens, GA: University of Georgia Press, 1986.

———. "Christianity and the Pursuit of Wealth." *Anglican Theological Review* 69, no. 4 (1987): 349–361.

———. *Earth, Sky, Gods & Mortals: Developing an Ecological Spirituality*. Mystic, CT: Twenty-Third Publications, 1990.

McDonald, Barry, and Patrick Laude, ed. *Music of the Sky: An Anthology of Spiritual Poetry*. Spiritual Classics Series. Bloomington, IN: World Wisdom, 2004.

McFague, Sallie. *The Body of God and Ecological Theology*. Minneapolis: Fortress Press, 1993.

———. *Life Abundant: Rethinking Theology and Economy for a Planet in Peril*. Minneapolis: Fortress Press, 2001.

———. *Super Natural Christians: How We Should Love Nature*. Minneapolis: Fortress Press, 1997.

McKibben, Bill. *Hope Human and Wild: True Stories of Living Lightly on the Earth*. St. Paul, MN: Ruminator Books, 1995.

McNeill, J. R. *Something New under the Sun: Environmental History of the Twentieth Century World*. New York: W.W. Norton, 2000.

Meadows, Donella H., et al. *The Limits to Growth*. New York: Universe Books, 1974.

Merchant, Carolyn. *The Death of Nature: Women, Ecology and the Scientific Revolution*. New York: Harper & Row, 1989.

———. *American Environmental History: An Introduction*. New York: Columbia University Press, 2007.

Miller, Timothy. "Hippies." In *Encyclopedia of Religion and Nature*, ed. Bron R. Taylor, 1, 779–780. London: Thoemmes Continuum, 2005.

Mitchell, Julia Benton. "Feminist Theologians and the Liberal Political Issues." In *Political Role of Religion in the United States*, 325–336. Boulder, CO: Westview Press, 1986.

Moe-Lobeda, Cynthia D. *Healing a Broken World: Globalization and God*. Minneapolis: Fortress Press, 2002.

Montefiore, Hugh. *Man and Nature*. London: Collins, 1975.

Moore, Mary Elizabeth. *Ministering with the Earth*. St. Louis: Chalice Press, 1998.

Moyers, Bill. *Moyers on America: Is God Green?* Video. Princeton, NJ: Films Media Group, 2006.

Muratore, Stephan. "All Creation Rejoices: Christian Femininity in Environmental Healing." *Epiphany* 7, no. 1 (1986): 26–35.

———. "Christian Ecology." *Epiphany* 8 (1988): 2–63.

———. "Earth Stewardship: Radical Deep Ecology of Patristic Christianity." *Epiphany* 10 (1990): 121–133.

———. "Ecumenical Ecology: The North American Conference on Christianity and Ecology." *Epiphany* 8, no. 1 (1987): 66–72.

———. "Holy Weakness, Strength from God: From Despair to Christian Ecology." *Epiphany* 6, no. 1 (1985): 74–77.

———. "Keeping Afloat: Stewardship in Machines, Money, and Farms." *Epiphany* 8, no. 1 (1987): 1–73.

———. "Reconciliation with the Environment, an Estranged Realm of the Spirit." *Epiphany* 6, no. 1 (1985): 84–88.

———. "Stewardship Is Enough: Ecology as Inner Priesthood." *Epiphany* 6, no. 1 (1985): 26–34.

———. "Taste the Fountain of Immortality: The Mystery and the Power of the Holy Eucharist." *Epiphany* 9 (1989): 6–47.

———. "Where Are the Christians: A Call to the Church." *Epiphany* 6, no. 1 (1985): 7.

Muir, John. *My First Summer in the Sierra*. Boston: Houghton-Mifflin, 1911.

———. *Our National Parks*. Boston: Houghton-Mifflin, 1901.

———. *Ramblings of a Botanist among the Plants and Climates of California*. Los Angeles: Dawson's Book Shop, 1974.

———. *Yosemite*. New York: Doubleday, 1962.

Myers, Norman, and Jennifer Kent. *Perverse Subsidies*. Washington, DC: Island Press, 2001.

Myerson, George. *Ecology and the End of Postmodernity. Postmodern Encounters*, ed. Richard Appignanesi. Cambridge: Icon Books UK/US, 2001.

Nash, James A. *Loving Nature: Ecological Integrity and Christian Responsibility*. Nashville: Abingdon, 1991.

———. "Natural Law and Natural Rights," *Encyclopedia of Religion and Nature*, vol. 2, 1169–1171.

Nash, Roderick. "The Greening of Religion." In *This Sacred Earth: Religion, Nature, Environment*, ed. Roger S. Gottlieb. New York: Routledge, 1996.

Noh, Jong-Sun. "The Effects on Korea of Un-Ecological Theology." In *Liberating Life*, 125–136.

Northcott, Michael S. "From Environmental Utopianism to Parochial Ecology: Communities of Place and the Politics of Sustainability." *Ecotheology* 8 (2000): 71–85.

———. *The Environment and Christian Ethics*. New Studies in Christian Ethics, ed. Robin Gill. New York: Cambridge University Press, 1996.

Oduyoye, Mercy Ambo. *Introducing African Women's Theology. Introductions in Feminist Theology*, vol. 6, ed. Mary Grey, et al. Sheffield: Sheffield Academic Press, 2001.

Oeschlaeger, Max. *Caring for Creation: An Ecumenical Approach to the Ecological Crisis*. New Haven: Yale University Press, 1994.

Oliver, Harold H. "The Neglect and Recovery of Nature in Twentieth-Century Protestant Thought." *Journal of the American Academy of Religion* 60, no. 3 (1992): 382–383.

Oliver, Mary. "Have You Ever Tried to Enter the Long Black Branches?" In *West Wind: Poems and Prose Poems*, 61. Boston: Houghton Mifflin, 1997.

Our Common Future. New York: World Commission on Environment and Development, 198.

Owen, Catherine. *The Wrecks of Eden*. Toronto: Wolsak & Wynn, 2001.
Paris, Ginette. "Éco-Théologie." In *Religion et Culture au Québec*, 331–342. Montreal: Editions Fides, 1986.
Parsons, Howard L. "The Commitments of Jesus and Marx: Resources for the Challenge and Necessity of Cooperation." *Journal of Ecumenical Studies* 22, no. 3 (1985): 454–473.
Paterson, John H. "Ecology: Crisis and New Vision." *Christianity Today* 16 (February 18, 1972).
Pearce, Joseph. "The Education of E.F. Schumacher." *Godspy* 2003–2004. www.distributist.blogspot.com/2007/01/education-of-e-f-schumacher.html (accessed June 8, 2008).
Plants, Nicholas. "Lonergan and Taylor: A Critical Integration." *Method: Journal of Lonergan Studies* 19, no. 1 (2001): 143–172.
Primavesi, Anne. "Ecology and Christian Hierarchy." In *Women as Sacred Custodians of the Earth? Women, Spirituality and the Environment*, ed. Aliane Low and Soraya Tremayne, 121–139. New York: Berghahn Books, 2001.
———. *From Apocalypse to Genesis: Ecology, Feminism and Christianity*. Minneapolis: Fortress Press, 1991.
———. *Gaia's Gift: Earth, Ourselves and God after Copernicus*. London: Routledge, 2003.
———. *Making God Laugh: Human Arrogance and Ecological Humility*. Santa Rosa, CA: Polebridge Press, 2004.
———. *Sacred Gaia: Holistic Theology and Earth System Science*. London: Routledge, 2000.
Rasmussen, Larry. *Earth Community, Earth Ethics*. Geneva: WCC Publications, 1996.
———. "On Creation, on Growth." *Christianity and Crisis* 45 (1985): 473–476.
Ravid, Frederick. "Kebash: The Martial Commandment to Subdue the Earth." *Epiphany* 7, no. 2 (1987): 66–70.
Reeve, Ted, ed. *Moderator's Consultation on Faith and the Economy*. Pushing the Boundaries: Christian Action. Etobicoke, ON: United Church of Canada, 2000.
Reeves, Earl J. "Evangelical Christianity and the Ecological Crisis." In *The Cross and the Flag*, 181–201. Carol Stream, IL: Creation House, 1972.
Regeneration Project, Interfaith Power and Light. www.theregenerationproject.org (accessed August 28, 2009).
Reidel, Carl H. "Christianity and the Environmental Crisis." *Christianity Today* 15 (1971): 4–8.

Rigby, Kate. *Topographies of the Sacred: The Poetics of Place in European Romanticism*. Charlottesville: University of Virginia Press, 2004.

Roberts, W. Dayton. *Patching God's Garment: Environment and Mission in the 21st Century*. Monrovia: MARC and World Vision, 1994.

Robinson, Mark. *Biblical Applications of the Doctrine of Creation*. Dallas: Dallas Theological Seminary, 1973.

Roche, Douglas. "The Ethical Basis of Change: Religion and Politics and the Hubble Space Telescope." *Religious Studies and Theology* 9 (1989): 9–15.

Rochon, Lisa. "St. Gabriel's Church: Cityspace, Seeing the Light on Sheppard," *The Globe and Mail*, October 2, 2006: R4.

Rogers, Donald B. "Coyotes and Sheep: What the Exile Teaches the Church About Ecological Education." *Journal of Theology* 94 (1990): 7–19.

———. "Lessons from the Exile for the Instruction of the Church in Ecological Education." *Religious Education* 85 (1990): 412–423.

Roszak, Theodore. *The Voice of the Earth: An Exploration of Ecopsychology*. New York: Touchstone Books, 1990.

Royal, Robert. *The Virgin and the Dynamo: Use and Abuse of Religion in Environmental Debates*. Washington, DC: Ethics and Public Policy Center/Eerdmans, 1999.

Ruether, Rosemary Radford. "Critic's Corner: An Unexpected Tribute to the Theologian." *Theology Today* 27 (1970): 332–339.

———. "Women's Liberation and Reconciliation with the Earth." In *Women and the Word*, 31–38. Berkeley, 1972.

———. "Rich Nations/Poor Nations: Towards a Just World Order in the Era of Neocolonialism." In *Christian Spirituality in the United States*, 59–91. Villanova, PA: Villanova University Press, 1978.

———. *New Woman, New Earth: Sexist Ideologies & Human Liberation*. New York: Seabury Press, 1975.

———. *Sexism and God-Talk: Toward a Feminist Theology*. Boston: Beacon Press, 1983.

———. *Gaia and God: An Ecofeminist Theology of Earth Healing*. San Francisco: HarperCollins, 1992.

———. *Women Healing Earth: Third World Women on Ecology Feminism and Religion*. Maryknoll, NY: Orbis, 1996.

———. ed. *Gender, Ethnicity, and Religion: Views from the Other Side*. Minneapolis: Fortress Press, 2002.

———. *Integrating Ecofeminism, Globalization and World Religions*. Lanham, MD: Rowman & Littlefield, 2005.

Rupp, George E. "Commitment in a Pluralistic World." In *Religious Pluralism*, 214–226. Notre Dame: University of Notre Dame Press, 1984.
Russell, Colin A. *Cross-Currents: Interactions between Science and Faith*. Grand Rapids, MI: Eerdmans, 1985.
Rust, Eric Charles. *Nature—Garden or Desert? An Essay in Environmental Theology*. Waco, TX: Word Books, 1971.
Ryu, Tongshik. "Man in Nature: An Organic View." In *Human and the Holy*, 139–153. Maryknoll, NY: Orbis, 1978.
Sachs, Wolfgang. *Global Ecology: A New Arena of Political Conflict*. London: Zed Books, 1993.
Said, Edward W. *Beginnings: Intention and Method*. New York: Columbia University Press, 1985.
Salleh, Ariel. *Ecofeminism as Politics: Nature Marx and the Postmodern*. London: Zed Books, 1997.
Salomon, Jean-Jacques. "The 'Uncertain Quest': Mobilizing Science and Technology for Development." *Science and Public Policy* 22, no. 1 (1995): 9–18.
Sammells, Neil, and Richard Kerridge, eds. *Writing the Environment: Ecocriticism and Literature*. New York: Zed Books, 1998.
Samsonov, I. "Modern Ecological Crisis in the Light of the Bible and Christian Worldview." *Journal of the Moscow Patriarchate* 11 (1988): 40–45.
———. "The Problem of Man's Relations with Nature in Russian Religious Philosophy and Culture." *Journal of Moscow Patriarchate* 12 (1990): 48–51.
Santmire, H. Paul. *Creation and Nature: A Study of the Doctrine of Nature with Special Attention to Karl Barth's Doctrine of Creation*. Cambridge, MA: Harvard University, 1966.
———. *Brother Earth: Nature, God, and Ecology in Time of Crisis*. New York: T. Nelson, 1970.
———. *The Travail of Nature: The Ambiguous Promise of Christian Theology*. Minneapolis: Fortress Press, 1985.
———. *Nature Reborn: The Ecological and Cosmic Promise of Christian Theology*. Minneapolis: Fortress Press, 2000.
Scanlon, Leslie. "People of Faith Concerned About Ecology; Churches Going 'Green'." *The Presbyterian Outlook*, February 26, 2007: 8–9.
Scharper, Steven B., and Hillary Cunningham. *The Green Bible*. 2nd ed. New York: Lantern Books, 2002. 1st ed. Maryknoll, Orbis Press, 1993.

Schaefer, James. "Environmental Ethics from an Interdisciplinary Perspective: The Marquette Experience," *Worldviews: Environment, Culture, Religion* 8, no. 2–3, 2004: 336–352.

———. "Religious Studies and Environmental Concern." In *Encyclopedia of Religion and Nature*, ed. Bron R. Taylor, 1373–1379. New York: Thoemmes Continuum, 2005.

Schaeffer, Francis A. *Pollution and the Death of Man: The Christian View of Ecology*. London: Hodder and Stoughton, 1970.

Scharlemann, Martin H. "Ecology and Eschatology." *Concordia Journal* 2, no. 3 (1976): 126–128.

Schauffler, F. Marina. *Turning to the Earth: Stories of Ecological Conversion*. Charlottesville: University of Virginia Press, 2003.

Schramm, James P. Martin. "Population-Consumption Issues: The State of the Debate in Christian Ethics." In *Theology for Earth Community: A Field Guide*, 1996, 132–142.

Schumacher, E. F. *Small Is Beautiful: A Study of Economics as If People Mattered*. London: Sphere Books, 1973.

Schwarz, Hans. "Eschatological Dimension of Ecology." *Zygon* 9 (1974): 323–338.

Searle, Rick. *Population Growth, Resource Consumption, and the Environment: Seeking a Common Vision for a Troubled World*. Victoria: Centre for Studies in Religion and Society, 1995.

Seibert, Ute, and Josefina Hurtado. "Conspirando: Women "Breathing Together." *Ecumenical Review* 53, no. 1 (2001): 90–93.

Shabecoff, Philip. *A Fierce Green Fire: The American Environmental Movement*. New York: Hill & Wang, 1993.

Shakio, Ronald G. "Religion, Politics and Environmental Concern: A Powerful Mix of Passions." *Social Science Quarterly* 68 (1987): 244–262.

Shinn, Roger Lincoln. "Survival Ethics: Toward a Zero-Growth Economy." *Christianity and Crisis* 32 (1972).

———. "The Energy Question: The Theological Issues." *Journal for Preachers* 4, no. 2 (1981): 4–8.

———. "Old Traditions, New Decisions: Protestant Perspectives." In *Contemporary Ethical Issues in the Jewish and Christian Traditions*, 53–77. Hoboken, NJ: Ktav Publishing House, 1986.

Shiva, Vandana. *Biopiracy: The Plunder of Nature and Knowledge*. Boston: South End Press, 1997.

———. *Stolen Harvest: The Hijacking of the Global Food Supply*. Boston: South End Press, 1999.

———. *Water Wars: Privatization, Pollution, and Profit*. Toronto: Between the Lines, 2002.

Sideris, Lisa H. *Environmental Ethics, Ecological Theology and Natural Selection.* New York: Columbia University Press, 2003.
Simon, Derek, and Don Schweitzer, eds. *Intersecting Voices: Critical Theologies in a Land of Diversity.* Ottawa: Novalis, 2004.
Sims, Bennette J. "History and Henry Ford." *Anglican and Episcopal History* 59, no. 3 (1990): 298–302.
———. "The North American Conference on Religion and Ecology: A Report to the Presiding Bishop." *Anglican and Episcopal History* 59, no. 4 (1990): 441–452.
Sittler, Joseph. *The Ecology of Faith: The New Situation in Preaching.* Philadelphia: Fortress Press, 1961.
———. *The Care of the Earth and Other University Sermons.* Philadelphia: Fortress Press, 1964.
———. "Ecological Commitment as Theological Responsibility." *Zygon* 5 (1970): 172–181.
———. *Essays on Nature and Grace.* Philadelphia: Fortress Press, 1972.
———. *The Structure of Christian Ethics.* Louisville: Westminster John Knox Press, 1998.
———. *Evocations of Grace: Writings on Ecology Theology and Ethics.* Grand Rapids, MI and Cambridge, UK: Eerdmans, 2000.
Smith, Douglas. "United Nations Conference on the Human Environment (1972) Stockholm, Sweden." In *Environmental Encyclopedia,* 854, ed. William P. Cunningham, et al. Detroit, MI: Gale Research, 1994.
Soelle, Dorothee, and Shirley A. Cloyes. *To Work and to Love: A Theology of Creation.* Philadelphia: Fortress Press, 1984.
Southgate, Christopher. *God, Humanity, and the Cosmos.* London: T&T Clark International, 2005.
Speth, James Gustave. *Protecting Our Environment: Toward a New Agenda.* Washington, DC: Center for National Policy, 1985.
———. "The Transition to a Sustainable Society." In *Proceedings of the National Academy of Sciences,* 89, 870–872, 1995.
St. John, Donald. "Ecological Prayer: Toward an Ecological Spirituality." *Encounter* 43 (1982): 337–348.
———. "Energy, Vision, and the Future: A Hopeful Reflection." *Anima* 9 (1982): 63–68.
Starhawk. *Webs of Power: Notes from the Global Uprising.* Gabriola Island, Canada: New Society Publishers, 2002.
Stone, Glenn C. *A New Ethic for a New Earth.* New York: Friendship Press for the Faith-Man-Nature Group and the Section on Stew-

ardship and Benevolence of the National World Council of Churches, 1971.

Strauss, James. *The Theology of Nature and the Ecological Crisis*. Eden Theological Seminary, D.Min thesis, 1974.

Streiffert, Kristi G. "The Earth Groans, and Christians Are Listening." *Christianity Today* 33 (1989): 38–40.

Swimme, Brian, and Thomas Berry. *The Universe Story*. San Francisco: HarperCollins, 1992.

Sykes, James T. "Three Takes on Late-Life Crises," *The Gerontologist* 42, no. 1 (2002): 140–141.

Tarnas, Richard. *Cosmos and Psyche: Intimations of a New World View*. New York: Penguin Group, 2006.

Taylor, Charles. *Sources of the Self: The Making of the Modern Identity*. Cambridge, MA: Harvard University Press, 1989.

———. *The Ethics of Authenticity*. Cambridge, MA: Harvard University Press, 1991.

———. *Modern Social Imaginaries*. Durham, NC: Duke University Press, 2004.

Taylor, Sarah McFarland. *Green Sisters: A Spiritual Ecology*. Cambridge, MA: Harvard University Press, 2007.

Teilhard de Chardin, Pierre. *The Phenomenon of Man*[sic], trans. Bernard Wall. London: William Collins Sons; New York: Harper & Row, 1959.

The Global 2000 Report to the President - Entering the Twenty-First Century. Washington: U.S. Council on Environmental Quality, 1980.

Thoreau, Henry David. *The Maine Woods*. Boston: Ticknor and Fields, 1864.

———. *Walden*. Boston: Ticknor and Fields, 1854.

Todd, Judith. "On Common Ground: Native American and Feminist Spirituality Approaches in the Struggle to Save Mother Earth." In *Politics of Women's Spirituality*, 430–445. Garden City, NY: Anchor Press, 1982.

Toolan, David. *At Home in the Cosmos*. Maryknoll, NY: Orbis, 2001.

Toulmin, Stephen. "Nature and Nature's God." *Journal of Religious Ethics* 13 (1985): 37–52.

Townes, Emilie M. "Womanist Theology: Dancing with a Twisted Hip." In *Introduction to Christian Theology: Contemporary North American Perspectives*, ed. Roger A. Badham. Louisville, KY: Westminster John Knox, 1998.

Tran Thu Phuong. "Gender Assessment in Natural Resource Use and Management and Environmental Protection in Vietnam: A Case

Study in Soc Son District." Saint Mary's University M.A. thesis, 2001.

Tsirpanlis, Constantine N. "Social and Political Dimensions of Eastern Orthodoxy." *Patristic and Byzantine Review* 3, no. 3 (1984): 215–222.

Tucker, Mary Evelyn, and John Grim, "Series Forward." In *Christianity and Ecology*, xv–xxxii.

UNEP. *Global Environmental Outlook 3*. Nairobi: United Nations Environment Programme, 2002, www.grid.unep.ch/geo/geo3/index.htm (accessed August 28, 2009).

Union of Concerned Scientists. "Warning to Humanity." *Renewable Resources Journal* 19, no. 2 (2001): 16–17.

U.S. *Congregational Life Survey Reports*, from the Hartford Institute for Religion Research, http://hirr.hartsem.edu (accessed August 28, 2009).

U.S. Department of Commerce, Economics and Statistics Administration, U.S. Census Bureau, www.census.gov/prod/2003pubs/c2kbr-31.pdf 1-10 (2003) (accessed August 28, 2009).

Van Dyke, Fred G. "Planetary Economies and Ecologies: The Christian World View and Recent Literature." *Perspective on Science and Religious Faith* 40 (1988): 66–71.

Viscusi, W. Kip, and Joni Hersch, "The Generational Divide in Support for Climate Change Policies: European Evidence." In *Harvard Law and Economics Discussion Paper No. 504 2005*, http://ssrn.com/abstract=670323 (accessed May 9, 2008).

Vitousek, Peter M., Harold A. Mooney, Jane Lubchenco, and Jerry M. Melillo. "Human Domination of Earth's Ecosystems." *Science* 277, no. 5325 (1997): 494–499.

Wainwright, J. A., ed. *Every Grain of Sand: Canadian Perspectives on Ecology and the Environment*. Waterloo, ON: Wilfrid Laurier Press, 2004.

Wall, Derek. *Green History: A Reader in Environmental Literature, Philosophy, and Politics*. London and New York: Routledge, 1994.

Ward, Barbara, and Rene Dubos. *Only One Earth: The Care and Maintenance of a Small Planet*. New York: W.W. Norton, 1972.

Warren, Karen J. *Ecofeminist Philosophy: A Western Perspective on What It Is and Why It Matters*. Lanham, MD: Rowman & Littlefield, 2000.

Weber, Leonard. "Land Use Ethics: The Social Responsibility of Ownership." In *Theology of the Land*, 13–39, eds. Bernard Evans and Gregory D. Cusack. Collegeville, MN: Liturgical Press, 1987.

Weston, William (Beau), "Gruntled Center: Faith and Family for Centrists, Sociobiology 4: The Grandmother Hypothesis" http://gruntledcenter.blogspot.com/2006/02/sociobiology-4-grandmother-hypothesis.html (accessed, August 28, 2009).

Westra, Laura, and Bill E. Lawson, eds. *Faces of Environmental Racism: Confronting Issues of Global Justice.* Lanham, MD: Rowman & Littlefield, 2001.

Weyler, Rex. *Greenpeace: How a Group of Ecologists, Journalists and Visionaries Changed the World.* Vancouver, BC: Raincoast Books, 2004.

White, Rodney R. *North, South, and the Environmental Crisis.* Toronto: University of Toronto Press, 1993.

White, Vera K. *Healing and Defending God's Creation: Hands On! Practical Ideas for Congregations.* Louisville, KY: Office of Environmental Justice, Presbyterian Church USA, 1991.

Whitney, Kimberly. "Greening by Place: Sustaining Cultures, Ecologies, Communities." *The Journal of Women and Religion* 19, no. 2: 11–25.

Williamson, M. "The Impact of Space Technology on Society." In *International Symposium on Technology and Society*, 139–147. Glasgow, UK: Glebe House, 1997.

Worldwatch Institute. *Vital Signs 2001: The Environmental Trends That Are Shaping Our Future.* New York: W.W. Norton, 2001.

World Resources Institute. *World Resources 2000–2001: People and Ecosystems.* New York: Oxford University Press, 2001, www.wri.org/publication/content/7970 (accessed August 28, 2009).

Wright, Robert. *Nonzero: The Logic of Human Destiny.* New York: Random House, 2001.

Young, Richard A. *Healing the Earth: A Theocentric Perspective of Environmental Problems and Their Solutions.* Nashville, TN: Broadman & Holman, 1994.

Zademach, Wieland. "The Freedom of the Truth: Marxist Salt for Christian Earth." *Occasional Papers on Religion in Eastern Europe* 7, no. 6 (1987): 7–33.

Zavestoski, Steve, "Constructing and Maintaining Ecological Identities: The Strategies of Deep Ecologists," In *Identity and the Natural Environment: The Psychological Significance of Nature*, 297–316, 2003.

INDEX

Allenby, Brad, 87–88, 136, 145
American Academy of Religion (AAR), 31
Anderson, Benedict, 3
Aquino, Maria Pilar, 99, 147, 148
Athabasca Oil Sands, 123–24, 154
Au Sable Institute, 86–87, 120

back to the land movements, 20, 26
Bakken, Peter, 43, 135, 137
Baum, Gregory, 58
Berry, Thomas, xi, 1, 12, 16, 19, 25, 29, 31, 32, 44–47, 58, 67, 72–74, 114, 122, 128, 132, 137, 138, 140, 141
Berry, Wendell, 40
Bingham, Sally, 116
Birch, Charles, 78, 81, 82, 139, 143
Birkeland, Janis, 91, 146
Bouchard, Bishop Luc, 123
Bradford, William 1, 2
Bramwell, Anna, 21–24, 131, 146
Burton-Christie, Douglas, 70, 142

Cannon, Robin, 58
Cameron, Anne, 92, 146
Carson, Rachel, 20, 22, 24, 25, 30, 35, 41, 45, 58, 133
Centre for Studies in Religion and Society, 86
Chavez, Cesar, 58
Chryssavgis, John, 69

Chung Hyun Kyung, 95, 96, 146, 147
Cizik, Richard, 119
Cobb, John Jr., 25, 29, 30, 35, 50–52, 77–78, 81–82, 108–09, 115–16, 135, 139, 141, 143, 151, 153
Columbia River Watershed, 123
Commoner, Barry, 20, 24, 25–26, 133
community: local, 58–59, 64, 89–104, 106, 108, 116, 120, 121, 122, 126; of discourse, 31, 34; of texts, 16–17, 68
congregations, xii, 106, 108, 112–17, 150, 153
Conradie, Ernst, 94, 147
Con-spirando, 96–97, 148
corporate power, 60, 149
corporations, 24, 60, 65, 99, 102–03, 140
cosmology, 33, 45, 58, 72–75, 77, 88, 100, 140, 143
Coward, Harold, 86, 128, 135, 141, 145
Creation Care, 119–20, 153

Daly, Herman, 51, 139, 141
Darwin, Charles, 75, 83, 144
Darwinian, 76, 81, 83
Davis, John Jefferson, 118, 153
Dawkins, Richard, 78
Deane-Drummond, Celia, 84–85, 145
Dennett, Daniel, 78
DeWitt, Calvin, 85–86, 145

Di Chiro, Giovanna, 58, 60, 140
Dietrich, Gabrielle, 98
drift, x, 15, 53, 55, 63, 66, 125, 131, 155
Dunn, Stephen, 32–34, 135, 139

Eaton, Heather, 102, 141, 147, 148, 149
ecofeminism, 33, 49, 89, 91–97, 102, 141, 144–49
economics, 24–25, 31, 45, 57, 62–63, 80, 82, 83, 89, 100, 138, 139, 151, 152
economy, 4, 6–7, 9, 12, 51, 69, 103, 126, 139, 141, 144
ecowomanism, 94–95
Edwards, Denis, 80, 144
Environmental Defense Fund, 58
environmental racism, 59–62, 90, 140, 147

Forum on Religion and Ecology (FORE), 31, 47

Gaia, 49, 74–77, 105, 138, 143, 150
Gebara, Ivone, 80, 97, 145, 148
Genesis Farm, 47, 122
globalization, 57, 60, 63–66, 99, 101–03, 141, 146, 147, 148, 149, 150, 156
Gnanadason, Aruna, 98–99, 102, 148, 149
Grandmother Hypothesis, 105, 109–12, 151
Greenpeace, 20–21, 27, 132, 133, 134
Green Sisters, 47, 122, 138
grief, viii, ix, 14
Griffin, David Ray, 51
Grim, John, 31, 47, 68

Habel, Norman, 66, 117, 141, 142
Hallman, David, 69, 142, 146
Haught, John, 78–79, 144
Heller, Chaia, viii, 49, 128, 138
Hessel, Dieter T., 67, 80, 141, 142, 144, 149

Holy Cross Centre, 31, 123
hooks, bell, 95, 147
hope: for the future, 1, 25, 41, 51, 74, 79, 107; green, 112, 114, 115 meaning in this book, i–x; practices of, viii, ix, xi, xiv, 34, 52, 56, 89, 99, 103, 105–24; social imaginary and, 3, 74, 125, 126; theological interpretation, 14–16, 41, 52; voices of, 98, 104; vision of, 16–17

International Monetary Fund, 64, 102
Interfaith Power and Light, 115–16, 120, 152, 153
Isasi-Diaz, Ada Maria, 90, 146
Ivakhiv, Adrian, 100, 101, 149

Jaspers, Karl, 13, 130
Johnson, Elizabeth, x, 68, 128, 142, 144
Johnson, Patricia-Anne, 94, 147

KAIROS, 117, 154
Kaufman, Gordon, 68, 142
Kearney, John, 101–02, 149
Keller, Catherine, 61, 141, 142
Kerans, Patrick, 101–02, 149
King, Carolyn, 83, 106–08, 109, 119, 130, 145, 150, 151

Latour, Bruno, 12, 130
Leiss, William, 47, 138
Lester, Rita, 93, 147
Lonergan, Bernard, 15–16, 27, 53, 55–57, 63, 66–68, 78, 131, 132, 134, 139; common sense, 15, 55–57; longer cycle of decline, 15–16, 55, 61, 63, 66–69
Lorentzen, Lois, 102, 141, 147, 148, 149
Lovelock, James, 75
Lummaa, Virpi, 111, 151

MacGillis, Miriam, 47, 122
Marsh, George Perkins, 19

Index

Martin, Julie, 94, 147
Marx, Karl, 5, 146
McDaniel, Jay, 78, 143
McFague, Sallie, 61–63, 68, 79–80, 81–82, 98, 141, 142, 144, 148
McKibben, Bill, vii, 16, 127, 132
Merchant, Carolyn, 12, 57, 92, 131, 132, 140, 146
Miller-Travis, Vernice, 89, 90–91, 145, 146
modernity, x, 1–3, 4, 5, 9, 12, 17, 40, 43, 45, 46, 49, 128, 131
Moe-Lobeda, Cynthia, 104, 150
Moltmann, Jurgen, 81
Moyers, Bill, 118, 153
Muir, John, 19, 132
mujerista theology, 90, 95, 146
Myerson, George, 36, 136

Nash, James A., 80, 130, 144
Nash, Roderick, 12, 130
Northcott, Michael, 102–03, 149

Pollard, William, 74, 143
poverty, 2, 22, 26, 27, 35, 51, 54, 57, 62, 63–65, 111, 121

racism, 48, 57, 94. *See also* environmental racism
Regeneration Project, 116, 152, 153

Santmire, H. Paul, 29, 40–41, 135, 137
Schumacher, Fritz, 20, 24–25, 26, 45, 46, 138
Scientific Intellectual Movements (SIMs), 28–37, 134, 135
Shepard, Peggy M., 104, 132
Sideris, Lisa, 23, 81–83, 84, 133, 144, 145
Sierra Club, 58, 101
Sittler, Joseph, xi, 27, 29, 41–44, 55, 134, 135, 137
Smith, Adam, 5, 6–7, 63
social imaginary: key forms, 5–9, 12; and religion, 7, 9–11; social location, xiii, 89–104, 90, 126

Soelle, Dorothy, 97, 148
St. Gabriel's Church, 114, 152
St. Michael's College, Toronto, 32–34
St. Stephen's Cathedral, 113–14
Starhawk, 26–27, 134
structural adjustment, 64–65
Swimme, Brian, 45, 47, 72–73, 138

Tanner, Kathryn, 56, 139, 140
Tar Sands. *See* Athabasca Oil Sands
Taylor, Charles, ix–x, xii, xiii, 1–18, 39–40, 46, 54, 55, 62, 66, 71, 89, 125, 128, 129–30, 131, 134, 136, 143; on Christianity, 13–14; *See also* social imaginary
Teilhard de Chardin, Pierre, 27, 28, 46, 134, 138
texts: community of, 16–17, 68; defined, viii
The Earth Bible, 61, 66, 141, 142
Tucker, Mary Evelyn, 31, 47, 68, 142
Turner, Christopher, vii, 127

United Farm Workers, 58–59
utopias, 17

Vineyard Church, 118–19

Wallace, Mark, 68–69, 142
Weyler, Rex, 21, 132, 133, 134
White, Lynn, 12, 45, 47, 153
Whitehead, Alfred North, 50, 74, 76, 78, 143
Whitney, Kimberley, 99–101, 102, 146, 149
Wilson, E. O., 78
womanism, 48, 94, 95, 147. *See also* ecowomanism
World Bank, 60, 64, 102
World Council of Churches (WCC), 29, 43, 69, 95, 98, 102
World Trade Organization (WTO), 64, 65, 102

Young, Richard A., 61, 141